DEEP-FREEZE SECRETS

DEEP-FREEZE SECRETS

by

Charlotte Trevor

PAPERFRONTS

ELLIOT RIGHT WAY BOOKS
KINGSWOOD, SURREY, U.K.

Made and printed in Great Britain by
Cox & Wyman Ltd, London, Reading and Fakenham

CONTENTS

LIST OF ILLUSTRATIONS

Acknowledgement is made to Birds Eye Ltd. for the information and illustrations which they have kindly supplied for this book.

To my family

*With thanks for being willing guinea-pigs
in the days when I too was learning my way
around deep-freeze cooking.*

DEEP-FREEZE SECRETS

1

The Facts on Freezers

A decade ago, the deep-freeze was a kitchen luxury which only the rich could afford, and for which it seemed only the restaurateur could claim economic justification.

In the 1970s advances in freezer technology have brought these useful devices well within the budget of the average family. Changing social conditions have made the possession of deep-freeze facilities a must for the efficient modern housewife who wants to save time, labour and moncy.

There are already some millions of home freezers in the U.K. It is reckoned that, in ten years, there could be as many in this country as there are refrigerators today.

Does this seem a flight of commercial fantasy on the part of the freezer manufacturers? Reflect that in 1956 only 8 per cent of British homes had a 'fridge'. By 1965 the figure had risen to 44 per cent. Today between half and three quarters of the 18 million homes in the U.K. are equipped with refrigerators.

Statistics show that we are moving into an era when the refrigerator is considered an essential. Already most refrigerators on sale are equipped with 'star compartments' for the storage of frozen food.

Why, then should it be necessary further to clutter the already appliance-crowded kitchen with a special 'deep-freeze' unit? The answer is that, useful as they are as temporary reserves, star marked compartments can be used only for storage of pre-frozen food and are not designed to freeze fresh food. A 'one-star' compartment in which the normal temperature is −6°C (21°F) will hold most commercially frozen lines for up to a week. A two star compartment at −12°C (10°F) will keep them up to a month. Three star compartments at −18°C or 0°F will hold them up to three months.

These storage compartments are usually small in size compared with even the smallest deep-freeze units, and are *not* suitable for the freezing of home prepared foodstuffs for long-term storage.

For example, the smallest freezer currently available on the British market, and costing around £40, is a two cubic feet model. It will hold 40 lb. of frozen food, as compared with the few packets capacity of the average refrigerator frozen food storage compartment.

What is a Deep-freeze?

A deep-freeze is a cold chamber normally operating at the −18°C (0°F) temperature necessary to hold already frozen foods in perfect condition. By using the fast freeze switch, it is also capable of being operated continuously at a temperature colder than −18°C (0°F) to ensure that a load of fresh food is frozen as fast as possible and within 24 hours. Many home frozen foods can be stored for long periods without significant loss of nutrients, flavour or texture.

Even so, this temperature is well above the −34°C (−30°F) at which commercial quick freezing plants operate.

The true deep-freeze, as opposed to the conservator or the ordinary refrigerator, is capable of simultaneously freezing fresh food and storing already frozen items. It must be remembered, when assessing the size of freezer required, that the weight of food to be frozen in 24 hours should not exceed that recommended by the freezer manufacturer. When a batch of food is frozen it can be repacked in the storage space. Each cubic foot of storage space will hold about 20 lb. of packaged frozen foods.

The freezing of fresh food has been poetically likened to the sleep of the legendary 'Sleeping Beauty'. In freezer preservation the natural process of ageing, which would ultimately end in decomposition, is arrested by the drop in temperature.

Successfully frozen food, when cooked or thawed according to the manufacturers instructions or recipe directions, is indistinguishable in condition from when put in the fast freeze compartment. Once removed from the freezer it will, of course, again be subject to deterioration at normal rates.

The freezer can only preserve food in the state it finds it. Poor quality food can never be improved by freezing. Top quality food can, however, be spoiled if the few simple rules of freezing are not followed (see chapters on packing and preparing).

Do not therefore waste precious freezer space on storing any but top quality contents.

Who Needs a Freezer?

The advantages of a freezer to the country-based wife are obvious. Deep-freezing is the perfect way to preserve and store garden produce, such as strawberries or green peas, which come in spate over a short season only.

Preparing food for freezer storage is quicker and easier than traditional methods such as bottling. The end products are also likely to be more acceptable to the contemporary, calorie-conscious palate than the sugar-laden jams and jellies of the 'Mrs. Beeton' style of preserving.

The range of farm, garden and countryside produce which can be freezer-stored is far wider than could be preserved by any other means. For example, freezer owners can virtually forget the phrase 'close season' and enjoy pheasant or salmon the year around.

It is not only farmers' wives who can benefit from buying a freezer. The *working wife* with a freezer becomes one who can compete in the elegant entertaining stakes with those who are at home all day.

There is an almost endless variety of party dishes, pleasing to both children and grown-ups, which can be prepared days or even weeks in advance of the event. These dishes only need thawing out and placing on the buffet table. The hostess is enabled to steal more time on a tight schedule to beautify herself for her guests.

Guests who just 'drop in' can always be genuinely welcome in the homes of freezer owners. Instead of having to take 'pot-luck' they can be pampered by being offered a choice of joint, poultry or fish.

The deep freeze can be a boon to the *bachelor* or *bachelor girl*. It can provide an ever-present backstop against the days when its owner is 'kept late at the office' until after the shops are shut.

We tend, as yet, in this country, to think of the freezer primarily as a good buy for the big family but it can be just as much help to the person living alone, especially if he or she is elderly. It is difficult, often embarrassingly so, to persuade the butcher, fishmonger or baker to cut off just enough for one. The kinds of cuts specifically set aside as small portion lots can be tasteless compared with larger joints. The person living alone who owns a freezer can save money by buying a bigger joint and cutting, packing and freezing individually in mealtime portions which can then be used as needed.

A freezer could be the answer for the elderly person who wants to remain independent but who cannot be bothered or perhaps is unable to cook elaborate meals for himself. Younger relatives may provide a balanced diet by stocking the freezer with cooked dishes which only need reheating over a ring.

The wife alone during the day may be edged into obesity because it is easier to nibble a bun or a biscuit than to cook a proper meal for one. She can be induced into more sensible eating habits if, when cooking a big meal for the family, she also plates up individual 'on a tray' luncheons for herself.

Slimmers, and anyone on a special diet, can have their meals set aside in special freezer cartons.

Diabetics and slimmers can freeze berry fruit without sugar and sweeten it, after thawing, with an artificial sweetener. Sugar is not essential when preserving fruit by freezing, but flavour and colour are better when it is used.

Fruit or vegetable purées for babies can be frozen in portion size containers and stored in the freezer for use when needed. Conditions of utmost hygiene must be observed and purées must be well boiled, cooled quickly and frozen immediately.

Ice lollies for *children* can be made with real fruit juices. They will cost less and may cause less damage to young teeth than do commercial types.

Older children and husbands can be left to fend for themselves for the occasional evening or weekend. There will no longer be any fear that they will subsist entirely on fish and chips or baked beans.

Anyone in the family who has a 'fad' for a particular type of food, can now have his little fancies indulged. When a dish comes up that he doesn't like, the freezer stock affords an alternative. You can prepare individual portions and label them for the potential consumer.

This freedom of choice is probably a reason for the comment by freezer owners that the family eats better, enjoys a wider variety of food, and has luxury items more regularly than before.

The freezer is invaluable to *anyone who has to watch the pennies*. It enables the housewife to take advantage of glut or special offer prices, in quantity. Surpluses can be stored to reappear at a later stage in the menu-planning cycle. Nothing is wasted. The food boredom when the Christmas turkey carcass makes its seventh appearance in a week becomes just a memory of the bad old pre-freezer era.

Chicken or turkey bones can be rendered into stocks and freezer-stored in pint or half-pint waxed board or polythene containers for later use in soups, sauces and risottos.

A recent survey conducted among freezer owners showed that most had initially bought the unit as a time and labour-saving device. In the final reckoning all had saved money too.

Two-thirds of the housewives questioned had saved up to £2 each month. Nearly one-tenth saved between £2 and £4 a month. The rest had saved much more!

How Much do Freezers Cost?
As with all consumer durables, much depends on size and type. The smallest British freezers from 1·75 to 2 cubic feet capacity cost around £60. A large capacity chest freezer could be anything between £150 and £200 with individual variations among the manufacturers.

It is possible to get freezers on hire-purchase terms. There are also schemes which offer freezers at discount prices as part of a deal also involving an annual, or longer, contract for the bulk supply of foodstuffs. (See chapter 5 on Bulk Purchasing.)

How Much Does it Cost to Run a Freezer?

Each cubic foot of freezer space normally uses between three and five units of electricity each week. The cost of running a 6 cu. ft. freezer (the common size for a small household) would be around 48p per week.

The running cost of a freezer is governed by several factors; how often it is used to freeze food, the frequency of door openings and where it is located being the most important. Siting in the coolest place in the kitchen and ensuring that the lid or door closes properly and is open only when necessary and keeping the interior free from frost and ice, as recommended by the manufacturer, will ensure the lowest running costs.

The running costs of a domestic freezer can be more than met by the savings made on housekeeping bills. For example, on staple commodities, such as green peas or fish fingers, purchase catering packs in quantity. You will find that, compared to buying the small domestic packs, you have reduced costs considerably. Typical of the sort of savings possible are 55p saved on 5 lb. of frozen peas, 50p on a sixty pack of fish fingers and 65p on a gallon of good quality ice-cream.

To get maximum saving on a freezer, have a reasonable turnover of food and use it in regular quantities rather than working for long-term storage.

How Big a Freezer Do I Need?

It is a common complaint among freezer owners to wish that they had been more adventurous on the matter of freezer size. Satisfied owners, when moving on to their second freezer, invariably buy a bigger one. Anyone who has learned the simple rules of successful freezer operation soon comes to regard the device as indispensable.

As a rule-of-thumb guide to the size of freezer needed allow 2 cubic foot, plus two cubic foot of freezer space for each member of the household.

What Sort of Freezer Should I Choose?
There are three basic types of freezer: the dive-in chest, which takes up a lot of floor space and is a broad box with a top opening lid; the reach-in upright with a front opening door and the refrigerator/freezer (or combination) which in one upright unit combines a refrigerator and a freezer.

Chest freezers are competitively priced and are slightly cheaper to run than the upright types. Whilst they can accommodate large bulky parcels of frozen food, they are less convenient to use and the contents are not readily accessible. The lid of a chest freezer cannot normally be used as a useful work top or as storage space.

Upright front opening freezers occupy the minimum floor space and are easy to load and unload. They need defrosting rather more frequently than chests but this is easier to do. Frozen food stored on shelves is easily accessible and the top of a small upright freezer can be used as a work top or as shelf space.

Refrigerator/freezer. This combination is becoming more popular because it combines a 2 door front opening refrigerator and a separate freezer. It can frequently stand on the same kitchen floor space as is needed for a refrigerator alone and has all the advantages of a self-contained upright freezer.

Other Points to Look For
Some of the newer freezers have a warning light to show if

they are not operating correctly. Almost all freezers have an indicator light to show when they are working. Battery-run warning lights or bells, which operate in the event of a breakdown, can be fitted inside the house for those models sited in a garage or outhouse.

Some freezers have a fast-freeze section, separate from the main storage chamber, for freezing fresh food.

Beware the conservator or frozen food storage cabinet

Do not buy a conservator if you really want a food freezer. Conservators are frequently seen in sweet shops where they are used for the storage of commercial ice-cream. They are made only to maintain temperatures of $-18°C$. They do not have sufficient heat extraction to hold the storage temperature at $-18°C$ ($0°F$) whilst freezing batches of fresh food.

If all you want is the means to hold commercially frozen food in bulk, a conservator could be a good buy, as it tends to be cheaper.

A new symbol was introduced in 1973 which distinguishes a genuine home freezer from a conservator. When using this new freezer symbol, the manufacturers of the freezer are obliged to state the weight of food which can be frozen in 24 hours.

When buying a second-hand freezer, check that spare parts and service are still available for that model. 'Elderly' bargains can often go wrong without hope of being repaired.

Where Do I Put a Freezer?

As the British home becomes increasingly mechanized, so the purchase of each new appliance creates a new space problem. However, a refrigerator/freezer can most easily be accommodated, often replacing an old refrigerator. An upright self-contained freezer will occupy the minimum floor space and is the next best type of freezer to consider. A chest freezer which occupies the greatest floor area per cubic foot of freezer space can be located away from the kitchen at the end of a passage, in an outhouse or garage, if there is no kitchen space available. –

If a freezer is sited where it could be visited by small children, without supervision, there is a danger that a curious toddler may climb inside. Forestall tragedy by seeking out a model which locks up, and keep the key out of reach of inquisitive small fingers.

Installation and Maintenance

After the freezer has been installed, switch on and set the dial at the lowest temperature possible. Leave it shut for at least six hours before packing with food, which should be done in stages.

Maintenance of a freezer is simple. The outsides of all freezers are easily kept spotless by a quick wipe with a damp cloth and a drop of liquid detergent. A little silicone polish can be used occasionally, if needed.

Inside is more important. Defrosting should be carried out two or three times a year with chest models, and more frequently with uprights.

Always study the manufacturer's instructions on how to defrost, as different models often require different methods.

Defrost when stocks are at their lowest. Remove and wrap the food preferably in aluminium foil or alternatively in several sheets of newspaper and store it in a refrigerator or cool place until it can be returned. Never attempt to re-freeze food which has completely thawed as it will lose its texture and flavour.

Spread a towel in the bottom of the cabinet to catch the ice and water. Scrape down the sides with a blunt spatula of wood or plastic (this is usually supplied as a tool with the machine). Never use a metal instrument.

Thawing can be speeded up by placing bowls of hot water inside the cabinet. Do not pour hot water directly into a chest cabinet, even if it is equipped with a drain, because this could damage the refrigeration system.

To clean the inside of the freezer use a solution of one tablespoon of bicarbonate of soda to half a gallon of warm water. Do not use soap, soapless detergents or any type of caustic cleaner. Finally, rinse with a small amount of clean, warm water and dry thoroughly. An hour after starting the motor, repack the freezer and replace the food. Set aside, for immediate use, any food which has begun to thaw. The whole defrosting operation should be completed in under two hours.

Moving House

When you move house, check whether or not the removers will be able to handle the cabinet loaded. If the move takes place within one day, your stock of frozen food should not suffer if you make sure the freezer is the last item on the van, the first out at the other end, and all ready to plug into the new socket. If the freezer must be unloaded, pack the contents in tea chests with 'dry ice', if available. Do not put 'dry ice' (even if wrapped) inside the freezer cabinet as it may damage the lining.

When Things Go Wrong

Freezer breakdowns are not as serious as many novice owners fear. For example, if there is a power failure, food within the freezer compartment should be all right for at least twelve hours, provided that the freezer lid or door is kept shut. It is rare for power failures to be of longer duration than this.

If the freezer stops working check that the wiring, plug and fuses are in order, before phoning the manufacturer's agent for service. It is wise to check about service facilities at the time of purchase. Keep the name and phone number of the service agent readily to hand – perhaps on a label on the side of the freezer itself.

The address of the agent will probably be found on the instruction leaflet or guarantee card, issued with the machine. Failing this, try the yellow pages of the local phone book. The Electricity Board may also have a list of local service agents for the models it displays in its showrooms.

In the event of a breakdown, do not open the door of the freezer. Reduce the surrounding temperature, as much as possible, by keeping the room in which the freezer is sited well ventilated.

Leave the food in the freezer unless the serviceman needs it removed. If it must be taken out, wrap it in aluminium foil or in several layers of newspaper. Put the parcel inside one or more blankets, on a cold stone floor or in a cool place. The food should keep safely for about two hours, but must be checked for thawing before it is returned to the freezer.

The contents of a freezer can be insured against spoilage for about £3 per cent annually.

Do not turn off the electricity supply for long periods if you have a freezer. One family returned from holiday to find 250 lb. of decaying food in the freezer. The electricity had been switched off at the mains along with the water and gas.

If you are worried about the danger of leaving electricity on while you are away, have the freezer put on a separate circuit.

2

Filling the Freezer

The favourite frozen foods of the British family are green peas, green beans, fish fingers and ice-cream. These products make up the bulk of the 15 lb. of frozen food consumed annually by every man, woman and child in the U.K. The 75 lb. of frozen food eaten annually by the Americans covers a much wider variety.

These commercially frozen products may be excellent but it would be a pity to use a freezer to house them only.

Popular frozen products are nationally distributed to shops, large and small. So why not let the grocer store them and reserve precious domestic freezer space for more adventurous lines?

Most housewives who own freezers do progress steadily away from these 'top of the pops' items. They advance to cakes, pastry and mousse; they pass to the long-term storage of meat, game and poultry. Finally they prepare their own 'for the freezer specialities'.

To draw up a list of what every family should have in its freezer would be absurd. One might as well try to dictate what every family should have in its larder.

Family tastes differ as do their needs. Balanced buying for a freezer owner should include 'convenience foods' and home prepared time-savers. For example, you can purchase green peas in bulk, and prepare basic stocks or sauces at home. Also include reserves against crisis, such as a spare loaf, or out of season luxuries, like strawberries for Christmas or grouse for February.

Only trial and error can show the correct proportions of these fares for each household.

Once they get over their initial caution about what to put in a freezer, most owners go through a period at the other extreme. They freeze anything and everything to hand for the sake of novelty and to explore the potential of their machine.

Normally this 'new toy' stage in freezer ownership wears off. This machine ceases to be a status symbol and a novelty and becomes an indispensable cook's tool.

To make the best use of the freezer ask yourself, before storing, 'will the freezing of the commodity in question save: time, money or perhaps a food crisis?' It is only worth occupying precious freezer space and using electric power if you are freezing items which would be difficult to preserve by other methods. For example, it would be pointless for an 'onion-loving' family to store its supplies in a freezer when a string of onions keeps perfectly well.

Cabbage can be frozen, but it's available the year around so why bother? Sprouts have a limited season and are at peak quality only for a short time so they qualify for a place. This only applies to families who like sprouts or who have guests who like sprouts. Freezers filled with products which are never eaten are wasting precious space and electric power.

Fruit and vegetables selected for freezing should always be of prime quality.

It is most important that whole poultry, chicken, duck turkey, especially if it has been stuffed, should be thoroughly thawed out before cooking. This can take up to 48 hours, or longer, for a large bird being thawed in a refrigerator.

What Can I Freeze?

Almost any kind of food can be frozen, although some freeze better than others. It is probably best to memorize the products which fail the freezing test and, having eliminated these, you can experiment as widely as purse and inclinations allow.

Foods Which Won't Freeze

Salad Vegetables such as lettuce, tomatoes, cucumber, endive, onions, celery and radish will lose their crispness on thawing. This is because of their high water content.

Celery can be frozen if it is intended for use as a cooked vegetable.

Tomatoes can be stored if converted to purée.

Hard-boiled egg, aspic, meringue and *jelly* are likely to become tough if freezer stored.

Whole Bananas will turn an unattractive black if freezer stored. However, mashed bananas, if mixed with lemon juice to retain the colour, can be stored and used as a garnish for ice-cream.

Custard-type desserts or sweet sauces with a cornflour base tend to separate when frozen and then thawed.

Mayonnaise. Freezing is not recommended because it will curdle on thawing. Most other salad dressings can be frozen successfully.

Boiled Potatoes tend to become watery when frozen.

Cream Cheese freezes poorly alone. When mixed with whipped cream, as a cheese dip, or sandwich icing, it will store well.

Foods That Freeze Well

Meat, Game and Poultry. All kinds of fresh meat freeze

satisfactorily. Commercially-cured meats, such as hams or bacon, when frozen, will deteriorate earlier than fresh meats. This is due to the salt in the preparation which keeps much of the moisture in the meat from freezing. The maximum storage for prepared meats (hams, bacon, etc.) is up to 60 days. Uncooked beef and lamb will keep up to twelve months, pork up to nine months, mince offal and tripe two months. Roasting chickens and boiling fowls will keep up to twelve months, ducks six months, and poultry giblets up to three months.

Game should be hung to taste before being placed in the freezer.

When thawing frozen meats, allow three hours per pound for small roasting joints and four to five hours per pound for large roasts, if they are thawed out on a shelf in the refrigerator. At room temperature the thawing out period will be one hour per pound for all sizes of joint.

Eat the meat within twenty-four hours once it has been thawed.

Buying Meat for Freezing

Meat should be matured for the correct length of time before it is frozen. If you plan to buy direct from the slaughterhouse, let the butcher know when you will require the meat for freezing. He can then supply it at the correct stage.

Beef needs to be aged for approximately one week after killing, veal for about two days. Mutton should be aged for four days and lamb and pork for two days.

Store meat in joint-size packs which correspond to the size of joint you would normally buy as fresh meat for the family.

The less tender cuts such as chuck, shank, brisket, plate and neck should be boned before freezing to save freezer space.

Chops, steaks and rib roasts are improved in appearance at table by retaining the bone. These should only be boned if freezer space is very short. The sharp edges of boned joints should be trimmed before placing in the freezer, so that they will not puncture the wrapping paper.

When freezing several steaks or chops, interleave the meat with two thicknesses of paper or plastic. This will make the parcel easy to separate when required for use.

When cooking meat, from the frozen or partially frozen state (if there is not sufficient time to thaw out fully) allow extra cooking time at a slightly lower temperature. Stewing and braising accelerate thawing greatly, so only a little extra time is required to stew or braise solidly frozen meat.

To speed up the thawing of any meat, place it in the breeze of an electric fan.

Chicken

Chicken for freezing should be prepared and drawn in the usual manner. Chickens should not, however, be stuffed before freezing as this limits the storage life of the bird. A stuffed chicken is best eaten within one month. The chicken giblets should be wrapped separately and stored beside the bird.

Joints of Chicken take up less freezer space than whole birds. To joint a chicken for the freezer cut it in half through the breast bone and back bone. Cut each half into two or three pieces. Pack on an aluminium or fibre food tray and wrap finally with foil and polythene.

Game should be treated exactly the same as chicken, for freezing purposes.

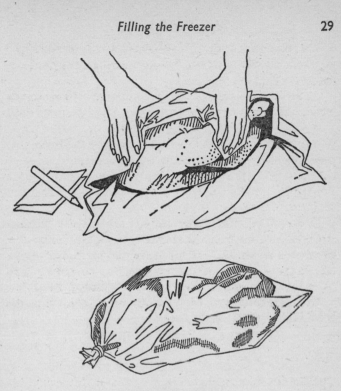

Fig. 1. Wrapping a Chicken.

Wrapping a chicken can be done with foil – as in the top sketch
or as in the lower picture the bird can be placed in a polybag and
secured with a clip.

Venison is handled, cut and packed in the same way as
beef.

Fish needs to be super-fresh (frozen on the day it is
caught) to freeze successfully. It should be scaled, gutted,
washed and dried. Although the heads and tails of fish will
freeze successfully, they are best removed.

To retain the 'just caught' appearance of fish, glaze

before the final freezing. To do this, freeze each fish individually and in its unwrapped state, until firm. Then immerse the fish in a bowl of icy cold, salt water several times. Between immersions, rest the fish on a wire tray. When it is completely coated with ice, wrap in polythene.

Whole salmon should be stuffed with paper before freezing, to retain their shape. They take approximately twenty-four hours to thaw out before cooking.

Dairy Produce

Eggs should never be frozen in their shells but as separate yolks and whites. When breaking eggs for freezer storage, remember that 16 large egg-whites will equal one pint of liquid. Twenty-four large egg yolks will fill a pint container.

Frozen eggs can be stored for six months. Thaw in a refrigerator or under cold running water and use in recipes in the same way as fresh eggs.

When storing egg yolks, add $\frac{1}{2}$ teaspoon of salt or one teaspoon of castor sugar to each two yolks. This will help to prevent coagulation. The packages of eggs should be labelled as 'sweet' or 'savoury'.

Two tablespoons of thawed white of egg approximately equal one egg-white and $\frac{1}{3}$ tablespoon of thawed yolk equals one egg yolk.

Milk

Ordinary milk does not freeze satisfactorily because the fat separates during the freezing and the milk becomes flaky. Homogenized milk can be stored for up to three months in the cartons in which it is purchased.

Cream

Use cream of at least 40 per cent butterfat content for freezer work. If it has not already been pasteurized, it should be heated to 160°F and cooled rapidly before being frozen. Sugar should be beaten into the cream before freezing and headspace allowed in the container. (See Chapter 3 on packaging for the freezer.)

Butter stores well in a freezer although its texture may become granular with long storage. Frozen butters may be stored up to six months.

Cheese. Soft cheeses freeze reasonably well for short periods if wrapped in foil. Camembert and similar cheese may be held for short periods to retard the ripening process. Hard cheeses, however, tend to become crumbly if stored for too long and need to be removed from the freezer at least twenty-four hours before they are required.

Vegetables

All vegetables intended for freezer storage should be blanched in boiling water. This retards the action of the chemical substances in vegetables called enzymes. These enzymes cause spoilage of both colour and flavour if the blanching process is carried out incorrectly.

Blanching

For the best results use a saucepan large enough to hold eight pints of fast boiling water with room to spare.

Put a maximum of 1 lb. of prepared vegetables into a wire basket or sieve and immerse in the fast boiling water for the specified time. (See charts on page 40—44.)

Time the process from the moment the water returns to a fast boil. The same water can be used several times.

Check times carefully. Underblanching will not destroy enzyme activity adequately and overblanching may spoil the texture of the vegetables.

As soon as blanching is complete, remove the basket from the water. Cool the vegetables quickly in a bowl of iced water or under a running tap. Drain thoroughly before packing.

All frozen vegetables can be cooked direct from the frozen state. The total cooking time should be reduced by the time originally taken to blanch them.

Fruit

There are three methods of preparing fruit for the freezer: dry, in sugar or syrup, or cooked and puréed. Select the method according to the type and quality of the fruit and the use for which it is required.

Fruit for freezing must be *just ripe*. After picking, it should be cooled and frozen as quickly as possible.

The Dry Method. Dry freezing is most suitable for soft fruits such as raspberries, loganberries, strawberries and currants. Prepare fruit in the normal way and pack dry or with sugar. Fruits in a sugar preparation should be coated in sugar and left standing for about ¾ of an hour. This gives them time to form their own syrup.

The Syrup Method is most suitable for fruits which require peeling and slicing and may become discoloured during preparation.

Syrups can be prepared with either 8, 13 or 20 oz. of sugar per pint of water. Choice will depend on personal preference and the natural sweetness of the fruit.

Puréed Fruits. Cook and sweeten fruit as if it were to be used directly for pie fillings or sauces.

Fruits to be served raw should be thawed slowly in an

unopened container, in the chill compartment of the refrigerator or in a cool place. Eat immediately after thawing.

Fruits to be cooked before serving should be heated slowly from their frozen state. (For further details on fruit preparation see details on pages 35–39.)

Bakery Products
Bread and Cakes should be wrapped and frozen as soon as they have cooled to hand temperature, after cooking.

Bought Bread can be frozen in its waxed paper wrapper and will last a week or more. Sliced bread can be toasted directly from its frozen state as a time saver.

Yeast Dough can be stored between baking sessions. Make up the bread dough in the usual way and put it into well-greased loaf-shaped aluminium foil containers. Freeze quickly and, when needed, bake in the usual way.

Bread Rolls or Fruit Buns should be made in the usual way and placed on flat trays which should be dusted over with fine semolina or ground rice. Freeze. When frozen hard lift from the trays and pack in layers in a suitable container. When required, replace on trays and bake in the normal way.

Sandwiches can be prepared in bulk for parties or packed lunches. Leave the crusts on the bread and pack the sandwiches tightly together in small parcels. Trim the crusts and cut to size before serving.

Cakes
Iced cakes may appear over-moist during the thawing out process. They will return to their normal state when thawing is completed. Butter-cream, iced cakes can be freezer stored satisfactorily and will keep for up to six months. Jam filling should be added to cakes after removal from the freezer. Wrap swiss rolls with paper between the coils

before placing in the freezer. This will facilitate filling the swiss rolls with jam. Cakes containing dairy cream are best sliced up while still frozen.

Pastry

Short, flaky, puff and choux pastry can all be successfully frozen. Pastry freezes best if it is frozen after mixing and before cooking. It has a freezer life of approximately six months.

Vol au vents or small bouchées are very useful to the hostess. Make the puff pastry in the usual way and place the vol au vents or small bouchées close together on layers of waxed paper in a suitable container, and freeze immediately. When needed, draw from stock and bake.

The puff pastry trimmings from the vol au vents may be pressed together into an oblong and rolled out thinly to be used to make sausage rolls. These sausage rolls can be packed in layers as before and frozen. Do not, however, put too many layers together until they are frozen hard, or the bottom layer will be squashed out of shape.

Quick reference chart on maximum keeping times for various types of food stuffs in freezer.
(Home frozen produce)

Raw Meat	*Months*	*Raw Fish*	*Months*
Bacon	1	Shellfish	1
Offal	3	Other fish	3–6
Ham	3–4		
Game, Veal,		*Vegetables*	8–12
Lamb	6–8	*Fruit*	6–9
Venison	8–10	*Cheese*	4–6
Poultry	10–12	*Ice-cream*	1
Beef	12	*Baked Goods*	1–2
		Precooked Meals	1–2

Preparation of vegetables and fruit for freezing

These charts detail the preparation needed, and blanching times where applicable, to freeze different fruits and vegetables for best results:

FRUIT	PREPARATION
Apples	Sliced. Peel, core and slice. Steam blanch for 2 minutes. Cool in iced water. Purée. Peel, core, and stew in minimum water, sweetened or unsweetened. Sieve and cool.
Apricots	Freeze whole, or sliced with stones removed. Best frozen in syrup.
Cherries	Remove stalks, stone, wash and dry.

FRUIT	PREPARATION
Gooseberries	Whole. Top and tail – wash and dry. Purée. Stew in minimum water, sweetened or unsweetened. Sieve and cool.
Grapefruit	Leave whole or segment before freezing. Can be sweetened.
Grapes	Remove stalks, halve, remove pips and freeze in syrup.
Greengages	Sweet. Pack dry or in sugar – 4 oz. per lb. Cooking. Syrup – $\frac{1}{2}$ lb. sugar to 1 pint water.

FRUIT	PREPARATION
Lemons	Leave whole or segment before freezing. Freeze peel for grating.
Melons	Dice or cut into balls. Freeze in syrup.
Oranges	Leave whole or segment before freezing. Freeze peel for grating.
Peaches	Freeze whole, or sliced with stones removed. Best frozen in syrup.
Pears	Freeze whole or sliced and cored. Best frozen in syrup.

FRUIT	PREPARATION
Pineapples	Dice before freezing. Can be sweetened first.
Plums	Freeze whole or sliced. Best frozen in syrup.
Rhubarb	Prepare in usual way, cook in minimum liquid with sugar added.
SOFT FRUIT Strawberries	Firm, dry fruit. Remove stalks. Dry pack or in sugar – 4 oz. per lb. Allow syrup to form before freezing.
Raspberries	Firm, dry fruit. Dry pack or in sugar – 4 oz. per lb. Allow syrup to form before freezing.

FRUIT	PREPARATION
Loganberries	Firm, dry fruit. Dry pack or in sugar – 4 oz. per lb. Allow syrup to form before freezing.
Blackberries	Firm, dry fruit. Dry pack or in sugar – 4 oz. per lb. Allow syrup to form before freezing.
Currants	Firm, dry fruit. Dry pack or in sugar – 4 oz. per lb. Allow syrup to form before freezing.

VEGETABLE	PREPARATION	BLANCHING TIME (MINS.)
Asparagus	Wash in ice-cold water. Cut into lengths to fit container. Do not tie into bunches. Blanch. Pack tips to stalks.	3–5
Beans, Broad	Shell, sort by size, blanch and pack.	3
Beans French or Runner	Select young, tender beans. Wash thoroughly. French – Trim ends and blanch. Runner – Slice thickly and blanch.	2–3 $1\frac{1}{2}$–2

VEGETABLE	PREPARATION	BLANCHING TIME (MINS.)
Broccoli	Select compact heads with tender stalks. Trim stalks evenly. Wash thoroughly. Drain well. Blanch. Pack tips to stalks.	$2\frac{1}{2}$–4
Brussels Sprouts	Select firm, tight, small, evenly sized sprouts. Remove outer leaves. Wash thoroughly. Drain well. Blanch.	4–6
Carrots	Select young carrots with good colour. Scrub, cut off tops, trim ends, wash well. Blanch. Freeze whole.	5–6
Cauliflower	Separate heads into sprigs. Wash thoroughly. Drain well. Blanch.	3–4

VEGETABLE	PREPARATION	BLANCHING TIME (MINS.)
Corn on the cob	Select young, yellow kernels. Remove husk and silks. Blanch.	6–10
Herbs	Chop or freeze in the sprig.	—
Leeks	Wash and slice. Blanch.	1
Mushrooms	Blanch in butter and oil mixture. Eliminate air from pack.	1 Oil Blanch
Onions – Fried Whole	Slice, flour and blanch in oil. Trim small button onions before blanching.	3 Oil Blanch

VEGETABLE	PREPARATION	BLANCHING TIME (MINS.)
Peas	Selected young peas, shell and sort by size. Blanch.	1–2
Potatoes	Cooked roast, duchesse or croquettes should be prepared in normal way.	—
Chips	Prepare as usual and blanch in oil.	4 Oil Blanch
Root Vegetables (e.g. Turnip, Swede)	Dice and blanch. Can also be frozen after full cooking and mashing with butter.	3

VEGETABLE	PREPARATION	BLANCHING TIME (MINS.)
Spinach	Select young leaves and trim. Wash well in running water. Drain. Blanch.	$1\frac{1}{2}$–2
Tomatoes	Slice and pack alone or as purée.	—

3

Packaging for Freezer Storage

Food destined for freezer storage should be carefully packed and wrapped. This will prevent the food drying out and losing its freshness and quality. It also reduces the number of times the freezer needs to be defrosted. Moisture escaping from the food adds to the frost on the sides of the freezer, which must be removed from time to time. Therefore, every item placed in a freezer should be protected by moisture- and vapour-proof materials.

Ordinary 'greaseproof' or waxed papers and gift wrap style cellophanes are not moisture- and vapour-proof and should therefore not be used.

When choosing wrapping, select strong materials that will not tear or puncture easily, but which will adhere to irregularly-shaped items like joints of meat.

Unless food is properly wrapped, it may suffer from the discoloured patches which are known as 'freezer burn'.

The polythene bag is probably the most useful tool of the freezer owner. It is cheap, strong and pliable, yet can be washed and re-used time and time again. Poly bags are easy to mould around meat and fish and suitable for fruit and vegetables.

Plastic- or paper-coated wire ties can be used to fasten polythene bags for the freezer. A special heat-sealing device,

Fig. 2. Materials suitable for packaging for the freezer.

Top; aluminium foil, foil dishes of various sizes. *Centre;* card cartons, polythene bags and metal closure strips. *Bottom left;* plastic boxes with lids. *Bottom right;* plastic film and freezer tape for sealing bags.

similar to those used in supermarkets, can be purchased. Bosch make an excellent domestic size heat-sealer which is also invaluable for lunch packs, gift wrapping, etc.

Waxed Cartons are made with a texture specially suitable for the freezer. They are produced by Frigicold Ltd., 166 Dukes Road, Western Avenue, London, W.3. Their ranges of freezer aids are widely stocked by stores.

When using waxed cartons, see that the lids fit well, otherwise drying out may take place. Line them with poly bags if the food is strongly coloured; this will prevent staining of the carton and any future contents.

Permanent freezer containers can be purchased in various shapes and sizes. They are made from either *rigid or flexible plastic. Aluminium dishes* are useful for pies, plate meals, puddings, etc.

Aluminium foil by the yard makes useful lids for cartons. Cut a piece large enough to fit down outside the rim of the carton and seal in place with freezer tape or a strong rubber band.

Freezer tape is a specialized version of the 'sello' self-adhesive tapes used for packing. It withstands low temperatures without losing its 'stick'.

A small hand suction pump can be purchased which will withdraw air from plastic bags. The plastic bag is clipped over the pump nozzle with an elastic band and the air is withdrawn. The bag is then slipped off the nozzle.

Air inside a package can cause discoloration and slow down the freezing process. Exceptions to this rule are stews, gravies and fruits in liquid. As it freezes the liquid will expand. Thus, in cartons or containers holding liquids for freezing allow a 'head space' of approximately half an inch for this expansion.

Tips on Freezer Packaging

When using plastic bags for food storage, try to shape the filled packages into a square or oblong. They will then take up less room in the freezer.

To package raw meat, press the plastic wrapping closely against the meat's surface, squeezing out all the air bubbles. Pack the meat flat. In the case of steak, cut in half rather than fold the meat. Interleave with polythene for easy separation when thawing out.

There are two basic methods of wrapping meat, baked goods and irregularly shaped foods: the flat fold wrap and the Butcher wrap.

The Flat Fold

1. Place the food diagonally close to one corner of a large sheet of wrapping material.

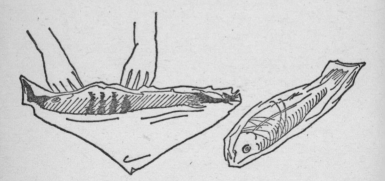

Fig. 3. The Flat Fold.
Diagonal wrapping is best for fish, which can be stored either whole – as shown – or with the head and tail removed.

2. Fold this corner up and over the food, then give the package a complete turn.
3. Bring the sides over the centre of the package and continue rolling to the opposite corner.

The Butcher Wrap

1. Place the food in the centre of the wrapping.
2. Bring the edges over several times until the fold is flat against the food.
3. Fold in the side edges and fasten the whole securely with freezer tape.

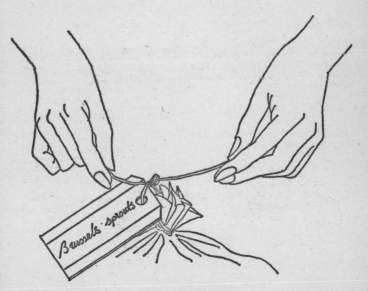

Fig. 4. Labelling.

Correct labelling is essential to find what is needed at speed from the inside of a well filled freezer.

These methods can also be applied to bread and cake wrapping.

All food destined for freezer storage should be labelled as well as correctly wrapped. (See sketch.) A wax or chinagraph pencil can be used to write directly on cartons or poly boxes. Other types of writing media dissolve at freezer temperatures.

Where labels are attached to polythene bags of food, use different coloured labels for different sorts of food. You can use perhaps red for meat, blue for fish, etc. This will make it easier to find what is required in a hurry. Packets of similar types of food can be stored in nylon shopping-bags of different colours. Or they can be placed in different layers of an upright or coffin type. Labels on frozen foods should give the date on which they were frozen and the quantity or weight of the package.

Most freezer owners, especially if their machine is a large one, keep a register of contents. This is to ensure that items are eaten in the correct rotation and while they are still in peak condition.

Tidiness in the freezer is an important time-saver. It will also help if you put new additions of a given food beneath or behind existing stocks of the same article. This makes sure that the older goods are used first.

When packing the freezer, do not over-fill but allow sufficient space for air circulation.

4

The Freezer as a Time-saver

Freezers, as well as saving shopping time, can be the cook's short cut to the preparation of dishes in the classic manner.

For example, your freezer should contain the following three basic sauces; *White flour-based Sauce* which is suitable for fish, chicken or vegetable dishes, soups or *vol au vent* fillings. *Tomato Sauce* which can be used for scampi provençale, spaghetti bolognaise or goulash, and *Basic Curry Sauce*. These time-saving sauces can be stored in $\frac{1}{4}$, $\frac{1}{2}$ or pint containers according to the size of the household.

Cooked rice can be deep frozen and used as an instant accompaniment to a variety of savoury dishes or as a savoury stuffing.

Odds and Ends of Cheese. Grate and freezer store, use to enliven sauces or to add the 'true' Italian touch to pasta preparations.

Chop and freeze fresh *herbs* in small quantities, or chop large bunches and place them in an electric liquidizer or blender, with a little water. Strain off most of the water and use the water as a soup flavouring to save it being wasted. Freeze the herbs in the refrigerator ice cube tray. When the herb cubes become solid, wrap them individually in foil. Each cube will be the equivalent of a tablespoonful.

Frozen cubes can be added directly as required to vegetables, sauces or casseroles.

A simple, fast way to prepare *parsley* for sprinkling is to
wash sprigs, then shake off surplus moisture and freeze the
sprays in poly bags. When required for use, take from the
freezer and crumble the frozen sprigs between the fingers
to 'chop' them.

Creamed cake mixture can be cooked in round, oblong
or patty tins and frozen without further preparation. Later
the different shapes can be used as follows:

> *Rounds* are filled with butter icing and decorated with
> glacé icing or fruit. They are also suitable for trifles or
> as a basis of a baked Alaska.
>
> *Oblongs* can be turned into small decorated cakes or
> Battenburg cake.
>
> *Patty tins* are decorated with glacé icing or butter icing.

Quick and pretty *cake or dessert decorations* can be made
by whipping cream with sugar. Pipe into rosettes on a
baking-sheet covered with silicone paper. Freeze the
trimmings as they are. Next day pack them into rigid
containers and store in the freezer. The finished rosettes
can then be popped on top of cakes and desserts, while
still frozen.

Whole Frozen lemons grate faster and more easily than
fresh ones. Slices of frozen lemon are marvellous in either
soft or alcoholic cold drinks.

Scraps of chocolate or nuts, suitable for cake or dessert,
can be stored in the freezer in small pots until required.
Left-overs of cold meat or fish can be turned into long-
keeping *potted spreads*. (Refer to recipe on page 119, con-
cerning potted spreads.)

Sausage rolls, for parties or the family, can be made well
in advance. Freeze them at the brushing-over stage in patty

tins or baking tray. The same process can be applied to *mince pies*.

The *last rose of summer* can be preserved for a birthday bouquet or a special dinner party table decoration. Cut, put in a poly bag and freeze for upwards of one month.

As an alternative to a tinned diet, make up the *Dog's Dinner* from bones, meat scraps or bought frozen dog food. Store in the freezer, seal it well and label it clearly so that Fido and not father helps himself to this tasty concoction. The same goes for *cats* too!

There is no need to waste the end of a loaf, if you are a freezer owner. *Breadcrumbs* stored at room temperature or even in a refrigerator soon develop mould. In a freezer they will keep almost indefinitely. Use the crumbs as they come out of the bag and let them thaw on the dish they are to top. Another saver is to freeze a bag of ready toasted or fried croutons for soup.

Iced coffee or *tea* can be made as instantly as hot. Freeze strong tea and coffee in ice-cube trays, wrap in foil and put in bags for storage.

Crumbled, cooked bacon bits can be frozen similarly to breadcrumbs. They can be used as an instant topping for baked jacket potatoes or any savoury dish. Combine with a moist ingredient, such as peanut butter, to make a delightful savoury sandwich spread.

Savoury butters are excellent time-savers for brightening all kinds of dishes. A prudent freezer owner will store:

Anchovy Butter Work 2 teaspoonfuls of anchovy essence into 4 oz. of butter.

Parsley Butter (sometimes known as maître d'hôtel butter) Work 2 tablespoonfuls of chopped parsley and some lemon juice into 4 oz. of softened butter.

Mustard Butter Mix 2 tablespoonfuls of French mustard into 4 oz. of softened butter.

Mint Butter Mixing 1 cup of mint leaves and 1 cup of parsley sprigs into 4 oz. of softened butter.

Freezers are invaluable for families who take *packed lunches to work or school.* Keep an assortment of ready-prepared sandwiches in the freezer. This can be an early morning time-saver. It also allows for impulse picnics or outings and can be a boon *for parties too.* When making sandwiches for freezer storage, spread the butter or margarine right to the edge of the slice as this will prevent the fillings from making the bread soggy. Jam sandwiches do not store well as the jam tends to soak into the bread as it thaws. Salad ingredients should not be used because salad vegetables lose their crisp quality when freezer-stored. The selection of possible fillings is almost endless. The following are especially successful:

Cottage Cheese with chopped gherkins.
Smoked, canned or *fresh salmon.*
Liver paté.
Danish Blue Cheese, mashed with chopped ham and softened butter.
Asparagus Rolls.
Minced beef and pickle.
Cottage cheese, chopped anchovy fillets and mustard.
Corned beef and mango chutney.
Peanut butter and chopped fried crispy bacon.

For sandwiches which are to be stored only for a day or so, waxed paper is sufficient wrapping. For longer periods freezer wrap is essential. The maximum number of sandwiches which can be packed together for good results are six to eight. Sandwiches should be allowed to thaw in their wrappings. If required for lunch, they should be removed from the freezer at breakfast-time. Sandwiches take eight hours to thaw in a refrigerator, or at room temperature, four hours.

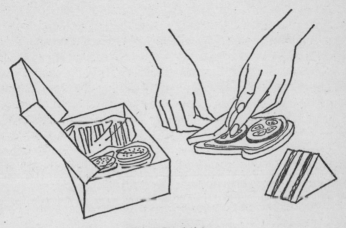

Fig. 5. Sandwiches.

Sandwiches for freezer storage are best packed in pairs. For time saving it is a good idea to prepare an entire packed meal in its take away box and store the whole in the freezer.

There is an advantage in using frozen sandwiches in a lunch pack which also contains fresh salad ingredients or crisp fruit. The frozen sandwiches will help to keep the fresh food crisp too.

When preparing sandwiches in bulk, note that one large loaf usually yields 20 to 30 slices depending on the thickness of the bread. Four ounces of butter and 10–12 oz. of filling will be needed.

Adult parties – especially *cocktail parties* – require a

selection of canapés, as well as sandwiches. Canapés are time-consuming to prepare so it helps if they can be made in advance and stored until needed. Hors d'oeuvre style preparations for parties should be stored individually according to type. Among the hors d'oeuvre/canapé ingredients which freeze well are:

Sliced sausage (including salami).

Sliced cold meats diced or cut in strips and dressed with oil and vinegar.

Smoked salmon, eel, buckling, sliced raw kipper.

Shrimps, prawns, potted shrimps.

Cooked cold vegetables with French dressing, such as cauliflower, beetroot, beans, raw sliced mushrooms.

Post party. Used cans of pickles and olives freeze well in their own liquor.

5
Bulk Buying

Families who rely on frozen foodstuffs as a regular part of their menu can save time and money by bulk buying.

Firms, specializing in the sale of produce in bulk, offer family size packs of popular items (like peas or fish-fingers) at prices far below the stores. One of these bulk purchasing schemes is operated by the Birds Eye Home Freezer Service. This is not a long-term contract and is flexible to the needs of each individual purchaser. (Other firms may operate similar schemes.)

Minimum order under this scheme at the time of writing is £5 worth of goods and the consumer must agree to be in to take delivery. Delivery men will not leave frozen foods on the doorstep. They insist on seeing the items safely stacked in the freezer, in order to ensure that their reputation for quality is not impaired by frozen lines being kept under poor conditions.

Users of the Birds Eye Service can also order the full range of Smethhurst's products (sister frozen food group serving the catering trade). This means that some items, like peas or frozen fish, are available at prices cheaper than those offered by Birds Eye. For the hostess who entertains regularly but is short on cooking time, Smethhurst's products can be a great advantage. She can chose specialities like Scampi Newburg or Duck à l'orange, which would normally only be available to restaurant owners. All these items could be stored in a conservator until required for eating, as they come already frozen.

Birds Eye have a nation-wide chain of depots, so this service is available to housewives anywhere in the U.K. For details of the scheme (delivery dates, etc.) check the yellow pages of the local phone directory under frozen food suppliers. Or you can look in the general directory under the trade name 'Birds Eye'.

Within the London area Harrods department store specialize in preparing cuts of meat specifically for the deep-freeze owner. They also supply a wide range of the exclusive American and Canadian speciality dishes for the freezer. These will add variety to the commercial food-stuffs usually stored.

The user of large quantities of commercially frozen food will find it profitable to deal with a local cash and carry grocery wholesaler. Such establishments would normally deal only with retailers and exclude domestic custom, as being unfair competition for their shopkeeper customers. The quantities required by the home-freezer owner sometimes puts him into the trade category, and cash and carry wholesalers are prepared to allow trade discounts.

There are also a number of companies which supply a freezer and keep it topped up on a contract basis, as a package deal. There are advantages and disadvantages to such a system and much will depend on the firm you deal with. The contract schemes deliver the produce at regular intervals and it may be found that the food is not sufficiently flexible to cope with personal idiosyncrasies.

The bulk buyer may find that the scheme will reduce his costs. A distributor who has the security of a certain outlet for his wares, can offer foodstuffs more cheaply than they are obtainable elsewhere. However, seasonal conditions may reverse this trend. The consumer may be forced to pay for his goods at the contractual, high price.

It is advisable to have experience of freezer stocking before deciding which scheme offers the most advantageous selections for your family. Yet these schemes seem angled to appeal chiefly to the novice freezer owner!

The purchase of a freezer, which, for a family size unit, can cost over £150, should only be made after family consultation and it should not be bought on impulse over the doorstep.

When shopping for freezers, visit the local electricity board showrooms to compare the various models on display. The Board's Home Economist staff can also give expert advice.

When using bulk purchased goods, think of new ways in which to vary them, otherwise the freezer, instead of reducing food boredom, can redouble it.

The following ideas may help to brighten bulk bought commercial vegetables:

Peas Mix with chopped skinned tomatoes or with spring onions, fried lightly in butter.

Beans Mix with crispy fried bacon bits and or chopped cooked onions.

Spinach Add a dash of horseradish sauce after cooking.

Sweet Corn Mix with a little double cream during cooking.

Broad Beans Serve in a cheese sauce.

As an alternative to saucepan boiling, cook frozen vegetables in foil parcels in the oven. Allow 30 minutes for peas and 40 minutes for other vegetables. Each parcel should contain a knob of butter and seasoning to taste.

Average Consumption	*Approximate Retail Price*	*Approximate Catering Pack Price*	*Approximate Saving*
1 lb. mixed Vegetables	18p per half pound	5 lb. pack 88p (17½p per lb.)	18½p
2 lb. Peas	15½p per half pound	5 lb. pack 95p (19p per lb.)	24p
2 lb. Beans	19p per half pound	5 lb. pack 89p (17·8p per lb.)	40½p
1 lb. Sprouts	19p per half pound	5 lb. pack 92p (18½p per lb.)	19½p
½ pint Ice-cream	12p per half pint	1 gallon £1·27 (8p per ½ pint	4p

By buying meat and fish in bulk, at least 75p per week can be saved (family of four).   75p

Savings on food bill for 1 week =   £1·81½p

This method can be used to save space on the top of the stove and fuel, when a joint is roasting.

Another way to save gas or electricity with frozen vegetables is to steam them over the potato saucepan. Allow five minutes longer than stated on the pack.

Frozen vegetables can be added in their frozen state to stews or casserole and will be done within the last twenty minutes of the main cooking time.

How Much Can I Save with my Freezer by Bulk Purchase?
Facing this page is an estimate of freezer bulk purchase savings working on the basis of a family of four using a chest type freezer costing around £100 and bought on easy terms of £40 down and the rest in instalments. On this basis the freezer not only paid for itself in around two years but additionally saved its owners £61.

6

Which Freezer is Right For My Family?

Today most of the big names in British refrigeration also produce deep-freezers. Many housewives, who already have a particular brand of refrigerator, will want to inspect the deep freeze of the same company.

The following firms are some of the makers or distributors of deep freeze units:

A.E.G. (Great Britain) Ltd.
Beekay New Era Ltd.
Bosch Ltd.
G.E.C.
Hotpoint
English Electric
Danfoss (London) Ltd.
Electrolux
Frigidaire
Hoover Ltd.
Kelvinator Ltd.
Lec Refrigeration Ltd.
Lo-Kold (Wholesale) Ltd.
Total Refrigeration Ltd.
Tricity.

Choice of a chest or upright freezer will depend on the

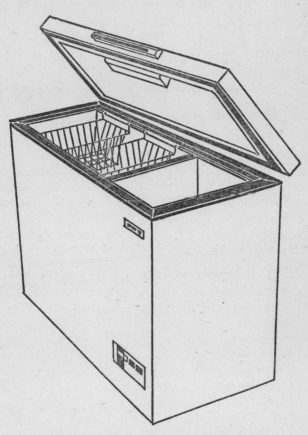

Fig. 6. Chest freezer with internal light fitting.

This makes it easier to see contents. The freezer shown here has
a deep compartment suitable for large joints of meat etc. and a
series of removable baskets for storing smaller packages.

space available for the freezer and the frequency with which it will be used. A family which has a cooking day once a week or once a month, and so uses its store of home frozen 'goodies' regularly, will probably find an upright freezer or refrigerator/freezer more convenient.

If you anticipate long-term storage, you will probably be well satisfied with a chest freezer. When buying a chest freezer remember that movable dividers for planned packing will enable you to make the best use of available space. Look out for a freezer with food baskets which won't rust and for a counter balanced lid which will stay open while you delve inside.

A freezer to be sited in the kitchen should have rollers which enable it to be moved out of its position when cleaning the floor.

Some upright freezers are now fitted with drop down doors on each shelf and this considerably reduces air spillage when the main door is opened. Other upright freezers and refrigerator/freezers are fitted with 'swing out' storage baskets for even easier access to the food. Many freezers are now fitted with an indicator light system to show when the power is switched on, when the fast freeze switch has been depressed and a warning light to indicate that the air temperature is warmer than normal.

Before you buy your freezer, think whether left or right opening doors are the most convenient. If the freezer is to be housed in a garage or outhouse which is not brilliantly illuminated, look for a model which has an interior light which goes on when the door is opened or the lid lifted.

Where pilfering may occur – such as in the small catering business – or where there might be a danger of children climbing inside, be sure to buy a freezer with a locking lid or door. Some freezers have a special fast freezing zone.

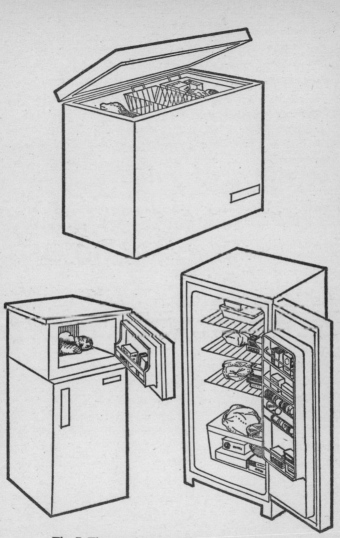

Fig. 7. Three main types of deep freeze cabinet:

Top; The chest freezer. *Bottom left;* Combination freezer and fridge unit, the deep freeze section rides on the top of the refrigerator cabinet. Although made in one piece each compartment has a separate door. *Bottom right;* Upright freezer.

This is useful if the freezer is to be used primarily to freeze your own garden produce or home-killed meats.

A number of freezers are supplied with a thermometer as a part of the equipment. Some have a lighted control panel to show when they are in action. Features such as these are useful for the novice owner. Whatever kind of freezer you choose, don't buy too small. The more food which can be stored, the more economical a home freezer becomes.

Prices of freezers vary between brands according to capacity, special fitments, etc. However, one would expect to pay £60–£85 for a 4 cubic foot chest type model to around £145 for an 8 cubic foot upright type.

The smallest free-standing models are of 1 to 2 cubic feet capacity. They are made to stand on top of either a working surface or a refrigerator and are most useful for the small household of one to three people, living in town.

If you are small in stature, check, before buying, that you can reach the bottom of the chest. Some chest freezers can be difficult for a small person to clean efficiently. A cleaning tool is usually sold with the freezer.

Any freezer operated by electricity will have to be sited near a power source. A 13 or 15 amp socket in a wall is the best plug-in point. Avoid fixing to a lighting socket, as this may overload the circuit and result in a power failure. Before using a freezer, ensure that the voltage indicated on the product corresponds with the mains voltage at home. If in doubt, check with the local electricity board. The freezer should be earthed.

Freezers can be fitted with two types of flexible cord conductors.

(1) If the flexible cord has green-yellow, blue and brown coloured conductors, the green-yellow is the earth core. Connect this to the pin in the plug, identified by the

letter E or by an earth symbol or coloured green or green-yellow. The blue core is neutral and must be connected to the pin identified by the letter N or coloured black. The brown core is live and must be connected to the pin identified by the letter L or coloured red.

(2) If the flexible cord has conductors coloured green or green-yellow, black and red, the green or green-yellow is the earth core. Connect this to the pin of the plug identified by the letter E or by the earth symbol or coloured green or green-yellow. The black core is neutral and must be connected to the pin identified by the letter N or coloured black. The red core is live and must be connected to the pin which is identified by the letter L or coloured red.

If the terminals of the plug are unmarked or you are in any doubt the only safe thing to do is to consult a qualified electrician.

7

Tips for Freezer Owners

When thawing *sauces* which have been frozen, always use gentle heat and preferably a double boiler. Break up the main block of the sauce as soon as some liquid begins to show at its base. As the sauce heats up, stir vigorously or apply a whisk to ensure smoothness. If extra cream or egg are to be added, they should be put in at this stage.

Seasoning may need adjustment before the sauce comes to table as many seasoning agents lose their potency during storage.

When making covered *pies* for freezer storage, omit the normal lid vents. Add these after the first ten minutes of baking the pie. Freezing pastry is especially successful. Indeed some cooks swear that pastry is the exception to the rule that nothing improves with freezing, as the cooling process seems to make some flaky pastries flakier.

Stews and soups intended for the freezer should have as much fat as possible skimmed off the top. It is primarily the fat turning rancid which limits the life of such freezer dishes. To save freezer space, make soups as concentrates. Use a smaller amount of water or omit the milk content in creamy soups. These can be added at a later stage when the dish is cooked ready to go to table.

When making thickened soups, substitute cornflour, rice flour, rolled oats or fine oatmeal for the plain flour content. This gives a better result especially if the soup is needed for long-term storage.

Where *cheeses*, such as Camembert or Valmeuse, are required for a special dinner party some days hence, freezing can be used to arrest ripening. If the cheese is at the perfect point for eating, it can be held in this state. However, such cheese must be given ample thawing out time, say two days in the refrigerator, and then time to get to room temperature. It will otherwise lack flavour and texture.

Among the excellent cheese dishes which can be made specifically for the freezer is *Iced Camembert*.

Take: 1 ripe Camembert
      1 demi sel or packaged cream cheese
      2 tablespoons hot milk
      3 tablespoons whipped cream
      salt and pepper to taste

Pass the rindless Camembert through a fine sieve then mix it with the cream cheese and hot milk until it becomes a smooth paste. Finally add the whipped cream and season before freezing in individual portions. On thawing, dust with paprika and serve garnished with watercress.

Fish to be served in cutlet form can be wrapped in individual portions in foil, together with salt, pepper, herbs, chopped lemon and a nut of butter. It can then be transferred directly to a baking-dish and baked while still in its parcel. This is most suitable for steaks of cod, salmon, hake or haddock.

Most *cakes* store well for longish periods but spicy cakes are an exception. A fortnight in the freezer is the maximum storage for this kind of cake. After this time changes in the flavour will become apparent on thawing.

Most packet directions on frozen *spinach* suggest using water to cook. However, both chopped and creamed commercial spinach can be cooked from the frozen state without adding water. Use a gentle heat and scrape the block of spinach with a wooden spoon until there is enough moisture in the pan to complete the cooking process.

To make a *quick cream of spinach soup* mix cooked spinach with a pint of white sauce (also good for freezer storage) and stir in a little cream just before the dish comes to table.

To make a *quick shrimp sauce* stir a jar of potted shrimps into a half pint of white sauce. The nutmeg and other seasoning in the potted product is quite sufficient flavouring to give the sauce a 'lift'.

Frozen *corn on the cob* is much improved in flavour if it is cooled in unsalted water and salt is added at the end. This method does not toughen the corn kernels.

If in a hurry to use a *frozen chicken*, place the bird (still in its bag) in cold water for an hour or so. Renew the water from time to time. Hot water should never be used.

When *assessing the quantities* of frozen food needed for individual servings, reckon that a packet of frozen fruit will contain from around 6½ oz. to 10 oz. The average single portion is about 3 oz. per person. Packs of frozen vegetables vary from distributor to distributor but most have a 10 to 12 oz. capacity sufficient for four medium-sized portions. Frozen fish normally comes in packs weighing between 14 and 16 oz. and will serve four.

When making *prepared dishes* for the freezer, undercook rather than overcook them. This is to allow for the added heating required to bring them back to their original condition. In most recipes reduce the heating time by approximately 30 minutes to save this problem.

Onion and garlic flavours may turn musty with long

refrigeration. In freezer recipes which require these seasonings, add a little before freezing, and then increase the strength of the flavouring at the re-cooking stage.

It is not always realized that pasta dishes, for example, spaghetti, macaroni and other Italian staples can be successfully frozen. To prepare pasta for freezer storage, slightly undercook the pasta in boiling salted water. Drain thoroughly and cool under running water in a sieve. Shake as dry as possible, pack into polythene bags and freeze.

Pancakes, either filled or plain store well. Make to your usual recipe and cool on a rack. Interleave with waxed paper to make separation simple and freeze in a polythene bag.

Professionally smooth *ice-cream* requires a special ice-cream machine, which will continue to whip the preparation as it cools. If a slightly rougher texture is acceptable to the family, reasonable ices can be made at home. Remember that too high a sugar content will inhibit the freezing and that ice-cream must be frozen quickly if it is not to develop a grainy texture. Flavourings should be as pure as possible since the strength of the taste will be reduced during freezing. Use vanilla pod rather than vanilla essence and genuine liqueurs rather than liqueur-like flavourings.

Eggs, gelatine and syrups will all help to give a smooth texture and prevent large ice crystals forming. The introduction of whipped egg-white gives a lightness to the finished ice-cream.

To make a *plain cream ice* suitable for the freezer take:
    one pint of thin cream
    one vanilla pod
    three ounces of sugar
    a pinch of salt.

Scald the cream with the vanilla pod, stir in the sugar and salt. Cool. Remove the vanilla pod and freeze mixture to a mush. Beat well in a chilled bowl, and continue freezing for 2 hours beating one more time before packing into containers and storing in freezer. (To scald – to heat a liquid to just below boiling point.)

8

Menu Planning

To use a freezer to maximum advantage, plan menus well in advance for the days when time saving is most necessary. Buy in bulk and have a 'cook-up' in which several days' meals can be prepared as one big job, instead of spacing the work out over a number of days.

The Americans, who have had freezers longer than ourselves, have brought this type of preparation to a fine art. They describe the process as 'Chain Cookery'. An example of Chain Cookery is to buy three chickens and prepare the following dishes from them: chicken stock, chicken pie filling, a made up chicken dish such as Chicken Tetrazzini or chicken curry, a pâté of chicken livers, a chicken pie, and some cold chicken sandwiches for packed lunches. Many of these dishes will lead into other menus.

It is not always advisable to depend on recipe books from the other side of the Atlantic to organize U.K. freezing programmes. In the States many of the basic ingredients which make up frozen menus are themselves deep frozen. This is because of the Americans' greater dependence on frozen food and the long distances between their market garden areas and the main urban usage points. Garden-style produce from California has to travel to places as far away as New York. It will arrive in a better condition if it is deep frozen or canned at its source rather than attempting the journey as 'fresh food'. It may, in a journey of several days, have to pass through great changes in climate such as we do not have here.

In the U.K. it is still possible, especially for a family which travels out by car into the country each weekend, to pick up fresh vegetables and home-killed meats. These can be brought back for home freezing. It would seem a pity not to take advantage of these country-fresh raw materials.

To make the most of menu planning with the deep-freeze, set aside a whole day for the chore of cooking for the freezer. It is certainly a help too if the cook can have a companion. Two or more women who own freezers can profitably get together on this task. They can share out the produce, the expense of the basic ingredients, and the work of converting into freezer specialities.

The husband and wife, who are both at work, can team together to make the freezer menus. This will cut the time spent preparing meals each evening in the week and they will have more time to spend together.

The notion of making the preparation of large quantities of food for the freezer into a team operation originated in the U.S.A. However, today it is probably of more importance in the U.K. For, in the American kitchen, machinery such as food mixers, dishwashers, etc., are the norm and the American housewife needs no additional help to prepare food on a large scale.

What is the most suitable food for freezer menu planning?

The answer is, the food which each household enjoys most and which is difficult to prepare in a hurry. For example, frozen chips are a favourite commercial freezer line but most still need to be fried when removed from their pack. They are not the most sensible item to freeze since, in most districts, fish and chips can be purchased ready to eat. More delicately flavoured lines or more exotic ones cannot. Anyway, chips have a high fat content and are more liable to go rancid than are fat-free items.

The experts can find ways of storing the few problem foods so that they can freeze almost anything successfully. But the novice may find it difficult to plate dinners which warm up as well as those she can buy ready-frozen. A great deal of scientific research is necessary to produce, for example, Yorkshire pudding batter which will cook in exactly the same time as partially pre-cooked slices of beef.

Concentrate, in the early stages of freezer ownership, on meals in which you know the main ingredients are good freezing lines. Fill in the gaps in the menu with other convenience foods which can easily be prepared fresh. For example, fish cakes only need re-heating in the frying pan and can be accompanied by commercial ready-to-cook chips, or by instant mashed potato.

When chain cooking for the freezer remember, a large batch of *mashed potatoes* can be used for plated dinners, the potato required in fish cakes and also for potato croquettes.

Beef stew will freeze well and last a long time if the fat is well trimmed from the cubes of beef and the stew is skimmed to remove surplus fat from the surface. Mashed carrots take up little space as they can be frozen in cuboid form and will only need re-heating.

Freezer stored *meat loaf* can be accompanied by fresh garden salads or given a hot topping of tomato sauce (also from the freezer) to turn it into an Italian style preparation. Add ready-to-serve spaghetti or rice, either as an instant product from the store cupboard, or as a freezer pack.

Frozen (empty) *sponge cake or pastry cases* can be brought out and filled with fresh garden fruit, tinned fruit

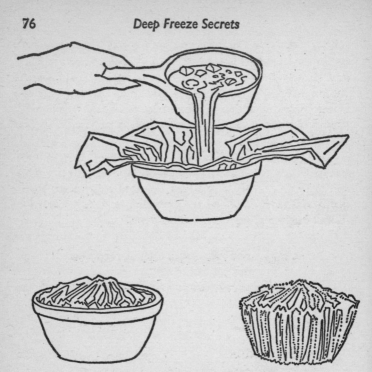

Fig. 8. The three stages in packing a stew or casserole dish.

First line the casserole base with foil and pour the stew into the foil lined base. Fold top over securely and stand in the freezer until set. Lift out the foil wrapped pack and store in the foil. When the stew is needed again simply unwrap the foil and place back in the casserole for reheating.

or preserves. Special purées or dry (sugar) packs of fruit from the freezer can also be used.

Put commercially bought *hamburgers or home-made meat rissoles* into their buns and toast direct from their frozen state to save time. Serve as a quick snack lunch, a supper, a children's party course or as a breakfast line at the end of an adult party, to counteract the drinks before the journey home.

Curry sauce can be re-heated and poured over a boiled egg in the time that it takes to cook the latter. Accompany it with sliced banana, sliced tomato and a fresh onion chopped fine and stood, for a few minutes, in a mix of lemon juice, sugar, salt and a sprinkling of dried rosemary. An oriental-type preparation with side dishes is ready in the twinkling of an eye.

Pour the same sauce piping hot over finely cubed lamb served with a dried bombay duck, straight from the pack, and a ready-to-use poppadum (bought from your local delicatessen). Pop under the grill and you have another Indian-inspired preparation in minutes.

6 Suggested Menus from the Deep-freeze

Combining Home Made and Bought In Items

Tomato Juice	Tomato Soup
Beef Casserole Broccoli Spears	Fish Cakes Peas Chipped Potatoes
Fruit Flan	
	Steamed Pudding
*	
Prawn Cocktail	*
Chicken Vol-au-Vent Potato Croquettes French Beans	Melon Balls in Syrup
	Scampi Newburg
Fruit Crumble	Lemon Mousse
*	*

Smoked Salmon   Pâté

Duckling à l'orange   Braised Steak
Asparagús   Cauliflower
Potato Croquettes
                                 Apple Turnovers
Chocolate Eclairs

9

Recipes:
Do-It-Yourself Style

Every cook who follows the simple rules of deep freezing, listed in the earlier chapters should quickly find ways to adapt her own favourite recipes to freezer storage. Her recipes may include steak and kidney, chicken or fruit pies, Victoria sponge, and other classics of English and Continental cookery.

The few recipes that follow are included because they are among the writer's favourite prepared freezer dishes.

Measures and Temperatures Table

Alternative weight, measure and temperature translations given in this book should cope with every cook's individual equipment. However, for personal favourite recipes the following guide may provide a useful translation key.

Oven Chart	°F	°C	Gas Setting
Very Slow	250–300	120–150	¼–1
Slow	300–350	150–177	2–4
Moderate	350–375	177–190	4–5
Medium Hot	375–400	190–205	5–6
Hot	400–425	205–218	6–7
Very Hot	450–500	232–316	8–10

All cup measures quoted in this book are American cup measures of 8 fluid ounces. U.K. standard measure cups take ½ pint (breakfast cup size) and ⅓ pint (tea cup size). There are 20 fluid ounces in a pint. All spoon sizes quoted in this book are British spoon sizes and the following translations may be useful for those who prefer always to use scales:

Ingredient	*No. of Spoons*	*Weight*
Sifted flour	3 tablespoons	1 oz.
Granulated or castor sugar	2 tablespoons	1¼ oz.
Butter	2 tablespoons	1¼ oz.
Cornflour	2 tablespoons	1 oz.
Rice	2 tablespoons	1¼ oz.

CAKES AND TEA-TIME ITEMS
Éclairs
½ pint water
4 oz. cooking margarine (not soft)
pinch of salt
1 oz. sugar
8 oz. plain flour
5 to 6 eggs

Combine the margarine, water, salt and sugar in a saucepan. Bring to the boil. When the margarine melts, add all the flour. Cook over a low heat, beating until the mixture forms a ball, and leaves the sides of the pan. Remove from the heat and beat in the eggs, adding one at a time. Continue beating until shiny and smooth. The mixture should be soft but not runny and the last egg may not be needed.

Drop the mixture from a teaspoon or tablespoon on to a greased baking tray, leaving at least an inch space between each eclair. For the classic chocolate éclair, the mixture can be piped into strips four inches long by one inch wide or patted into shape with a pair of spoons. Bake for 30 minutes in an oven at 375°F (regulo 5). By this time the éclairs should be puffy, golden brown and wholly free from moisture. Remove from the oven, cut slits in the sides and return for a further 5 minutes baking. Cool on a cake rack, wrap and freeze. This should make about 24 éclairs.

To serve: Remove from wrappings and place in a 325°F (regulo 3) oven for 10 minutes. Cool and fill. As well as the classic tea-time filling of dairy cream, they can be filled with ice-cream or with any savoury filling as a cocktail savoury or hors d'oeuvre.

Bread Making with a Freezer

If you like to bake your own bread but find it is not always easy to buy fresh yeast, remember that yeast can be stored in the freezer.

Divide a large quantity of yeast into cubes of approximately 1 oz. weight each. Wrap these in polythene or foil and store in a plastic container in the freezer. Each 1 oz. cube of yeast will require about 30 minutes thawing at room temperature before it can be used for baking.

Unbaked bread dough can be stored in the freezer for up to two weeks. Prepare the dough in the usual way and allow to rise for the first time then shape and pack in freezer wrappings. Prevent the formation of a crust by brushing the dough with oil or melted butter before wrapping and storing.

Thaw bread dough in a warm room. The quicker it thaws the better the result will be. The second rising of the bread will take place during the thawing.

Basic Bread for Freezer Storage

3½ lb. white flour
3½ teaspoons salt
1 ounce yeast
1 teaspoon sugar
1¾ pint warm water

Mix the salt and flour together. Cream the yeast with the sugar and add to the warm water. Make a well in the centre of the flour and pour the liquid into the well. Sprinkle on, or mix in, a little of the flour to form a pool of batter and allow to stand in a warm place for 20 minutes. Mix to an elastic dough using more water if required. Knead well until the dough leaves the basin clean. Put to rise in a warm place until the dough has doubled its size. Shape and wrap as instructed.

Basic Biscuit Dough For Freezer Storage

2 eggs
1 lb. flour
8 oz. butter
8 oz. castor sugar
A little milk
1 level teaspoon baking powder.

Sieve the flour and baking powder onto a piece of paper. Cream together the butter and sugar until light and fluffy. Add the beaten eggs a little at a time, alternately with a few spoonfuls of sifted flour. Mix in the rest of the flour using sufficient milk to make a firm paste.

Cut into biscuit shapes, interleave with paper (to make them easy to separate) and freeze. Or freeze as a single chunk, to be cut in slices for subsequent baking. This is a basic recipe which can be converted to any flavour the family likes. Coffee essence or vanilla essence are especially good with this recipe and should be mixed in with milk. A teaspoon will be sufficient to flavour the whole batch of dough. Or split the quantity into several batches, each of which can be differently flavoured.

Scone Mixture for Freezing

1 teaspoonful salt
1 teaspoonful cream of tartar
3 cups of plain flour
2 oz. lard
1 teaspoonful bicarbonate of soda
A little milk

Sift together the salt, cream of tartar, flour and rub in the lard. Dissolve the bicarb in the milk and add to mix. Add more to give a dough of soft, pliable but not wet consistency which will keep its shape. Roll out on a floured board. Fold over. Wrap in a polythene bag. Freeze. Thaw in a warm place until the dough is once again at handling consistency. Roll, cut into rounds and bake.

DESSERTS

Bavarian Cream

4 egg yolks
½ pint milk
2 oz. sugar
¼ oz. gelatine

Heat the egg, yolks, milk, sugar and gelatine in a double boiler stirring with a wooden spoon until the mixture thickens. Cool in a basin of cold water and pour into individual moulds. Cover with foil lids. Freeze. Thaw overnight in a refrigerator.

Serve with whipped cream or with fruit.

Cherry Freeze

2 lb. of cherries, dark variety
Sugar to taste
Thin sliced white bread and butter

Stone the cherries and cook gently with the sugar. Allow to cool. Take a foil dish and layer the cherries in between slices of bread and butter. Cover with lid and freeze. During the freezing process the mass will become jelly-like. On thawing, serve with thick cream.

Coffee Delight

3 oz. castor sugar
4 oz. butter
4 oz. breadcrumbs
5 tablespoons black coffee

Cream butter and sugar. Mix in breadcrumbs and coffee. Press into foil dish. Cover with lid. Freeze. Thaw 45 minutes in refrigerator.

Cranberry and Orange Relish

1 lb. cranberries
2 oranges, quartered and seeded
2 cups of sugar

Chop the cranberries and the oranges together, mix with the sugar. Freeze.

Lemon Pudding

2 oz. cornflakes
3 eggs
4 oz. sugar
1 tablespoon grated lemon peel
3 tablespoons lemon juice
½ pint double cream

Crush cornflakes and sprinkle a little in six foil jelly cases. Beat egg whites to soft peaks and gradually beat in sugar until stiff peaks form. In another bowl, beat yolks until thick and beat in lemon peel and juice until well mixed. Whip the cream lightly, then fold egg yolk mixture and cream into egg whites until just mixed. Put mixture into cases and sprinkle with more cornflake crumbs. Add foil lids to cases – freeze.
To serve: Thaw in refrigerator for 30 minutes.

Rhubarb Crumble

1 lb. rhubarb, peeled and cut
6 oz. flour
3 oz sugar
4 oz. butter

Place the fruit in a buttered, foil dish. Rub the fat into the flour and sugar until the mixture has a crumbly consistency. Sprinkle over the fruit and press down firmly. Cover and freeze. To bake: place the still frozen dish in the oven and cook at 400°F (gas regulo 6) for 30 minutes. Cook for a further 45 minutes at 375°F (gas regulo 5).

This recipe can be adapted to all types of firm fruit and works especially well with apples, pears or firm plums.

Parfait

¼ cup water
1 cup sugar
2 egg whites
1/16 teaspoon salt
2 teaspoons vanilla
1 pint double cream

Cook the sugar and water on a high heat without stirring until it 'spins a thread'. Pour this slowly over the stiffly beaten egg whites to which the salt has been added. Continue to beat until thick. Cool, add the vanilla and combine with chilled, whipped cream. Freeze without further stirring.

To serve: Thaw to the consistency of ice cream. Put into parfait glasses, top with whipped cream and garnish with fruit.

This sweet can also be prepared with liqueurs for adult parties and crème de cacão is a particularly successful additive.

Fruit Fritters

Batter
Slices of fresh pineapple, apple or bananas sprinkled with
    lemon juice

Dip fruit in batter and fry until golden. Drain and pack. Freeze, remembering to interleave with paper so that individual portions will be easy to separate.

To serve: Remove from freezer and allow to thaw. Reheat in hot fat and serve.

POULTRY

Chicken Sticks

1 lb. cold chicken (cooked)
4 oz. cooked rice
Chicken stock
3 eggs (well beaten)
Crumbs
Seasoning
Fat for frying

Dice or mince the chicken meat and combine with the cooked rice. Add chicken stock and beaten eggs to moisten. Season liberally and shape into fingers. Roll in breadcrumbs. Pack into containers and freeze.

Roast Duck with Orange Sauce

One 5–6 lb. duck

Roast the duck in a 475°F (regulo 9) oven for 20 minutes. Pour off fat. Reduce heat to 350°F (regulo 4) and roast for a further hour and a half. Remove the duck – chill – freeze.

To reheat for dinner: Place duck still frozen on a rack in a shallow roasting pan. Roast again at 375°F (regulo 5) for a further 1½ hours.

The Sauce

1 cup sugar
2 cups redcurrant jelly (16 fluid oz.)
1 cup dry sherry (8 fluid oz.)
3 cups orange juice (24 fluid oz.)
3 tablespoons cornflour
1 cup water (8 fluid oz.)
Two peeled and quartered oranges

Combine the orange juice, redcurrant jelly and sugar. Cook over a low heat for 15 minutes. Stir in the sherry. Cook for a further 5 minutes. Blend the cornflour with the water and mix into the sauce until it thickens. Freeze. Before sending to table, add the two peeled and quartered oranges with the pith removed (alternatively tinned mandarins) to the sauce.

As an alternative preparation, the duck can be fully cooked – the meat neatly carved off the bones entirely – and placed in the sauce. This gives double service as the freezing liquor also.

Chicken Tetrazzini

2 lb. cooked chicken meat diced
12 oz. spaghetti cooked in chicken stock and drained
1 lb. mushrooms sliced
4 oz. butter
1 pint chicken stock
½ pint cream
Glass of sherry
3 oz. grated Parmesan

Fry the mushrooms gently in half the butter. Make a roux (the classic base for all gravies and sauces) of the remainder with the flour, add the stock. Bring to boil. Stir in the sherry, cream, seasoning and cheese. Pour a little sauce over the chicken using the remainder to blend with the spaghetti. Arrange the mix in a foil dish topping with the chicken. Cover and freeze.

Serve by baking in a moderate (350°F or regulo 4) oven for 1½ hours. Sprinkle the top with some further Parmesan cheese to brown gently 30 minutes before serving time.

Chicken Ramakins

This is a useful dish when you are chain-cooking chickens since it can be prepared from chicken off-cuts. It is intended for individual portion service and will be easy to store and ideal for a 'dinner alone'.

The substitution of some truffle for the mushrooms will turn this into a luxury dish.

6 oz. raw minced chicken
2 eggs
Seasoning to taste
2 mushrooms
2 tablespoonsful of lightly whipped cream
½ oz. butter

Add the egg yolks to the minced chicken making the mixture as smooth as possible. If you have an electric blender you can use this for the purpose, otherwise it may be necessary to sieve the mixture. Season. Add the mushrooms. Stir in the cream and fold in the beaten whites of eggs.

Turn the mixture into well buttered foil baking cases and freeze.

The dish will be fluffier and more soufflé-like if the mixture is frozen minus the cream and beaten egg whites. Add these on thawing.

Cook in a moderate to hot oven until well risen and browned. Sufficient for eight servings.

Chicken with Sour Cream

2 lb. of chicken – boned, skinned and cut into small pieces
¼ pint sour cream
Dash Paprika pepper
Dash Worcester sauce
1 dessertspoonful of lemon juice
Salt and pepper to taste
½ oz. butter
Breadcrumbs

Mix cream, lemon and seasoning. Cover the chicken with the mixture and coat in breadcrumbs. Place in a well buttered foil dish and bake in a moderate oven for approximately 45 minutes. Cool, wrap and freeze.

Serve by baking the chicken, in its foil dish, for approximately 30–45 minutes in a hot oven. If the fowl is required crisp, remove the foil lid of the dish ten minutes before serving.

SOUPS FOR THE FREEZER

Gaspacho (Spanish Style Cold Soup)

1½ pints canned tomato juice (or tomato juice from the freezer)
1 lb. cucumber
Pinch of pepper
Pinch dried garlic
2 tablespoonsful olive oil
4 tablespoons wine vinegar
A little sugar and salt according to taste
Chopped parsley according to taste

Peel the cucumber and grate it. Mix all the ingredients together using the parsley as a topping. Freeze.
To serve: Thaw at room temperature remembering that the soup is at its best when served ice cold.

Onion Soup (French Style)

1½ lb. onions
2 oz. butter
3 pints beef stock (from the freezer)
Salt and pepper
2 tablespoons cornflour

Slice the onions finely and cook gently in butter until soft and golden. Add stock and seasoning. Bring to the boil simmering for 20 minutes. Thicken with cornflour and simmer for 5 minutes.

To serve: Reheat in double boiler stirring gently. Serve with slices of toasted French bread on which butter and grated cheese have been melted during toasting.

Lettuce Soup

Lettuce cannot be frozen successfully but lettuce soup, which can be freezer stored, is a useful way of using up this salad crop when in glut. The same method can be used for a cauliflower soup, but the raw cauliflower must be sautéed first.

1 lb. lettuce (or cauliflower)
pepper and salt to taste
½ or ¾ pint stock
1 oz. butter
1 oz. flour
1 pint milk
1 egg yolk
A little cream

Boil the lettuce in the stock and purée. Make a roux of the flour and butter. Add the milk, pepper and salt to the vegetable purée. Combine the purée and the roux. Add the egg yolk and cream.

Many experts feel that this soup should not be reboiled after the egg and cream have been added and these ingredients do not freeze well, so it is best if they are added at the thaw and reheat stage.

Tomato Soup

2 lb. tomatoes
2 oz. mushrooms
1 sliced leek
2 sticks of celery
1 lemon
Salt and pepper to taste
3 pints stock
2 egg yolks
2 oz. ground rice
¼ pint single cream
Butter for frying

Fry all the vegetables, except the tomatoes, in butter until golden. Add the stock and simmer with the lemon juice and tomatoes. Sieve. Mix in the egg yolks, rice and cream. Cook slowly for a further 10 to 15 minutes without actually boiling. Season. Cool and freeze.

Cucumber Cream Soup

Cucumbers, like lettuces, will not freeze. However, in soup form they can be successfully freezer stored.

2 small cucumbers, diced
½ onion, chopped fine
1 tablespoonful spinach purée
2 pints white stock
Salt and pepper to taste
1 oz. butter
1 oz. flour

Sauté the onion and cucumber gently in a little butter, reserving the remainder to make a roux with the flour. Add the onion, cucumber and stock to the roux and simmer gently for a few minutes. Add the spinach and seasoning and place the mix in an electric blender or pass it through a sieve. Pack into cartons and freeze.

When reheating the soup, add either double or whipped cream and some cut chives or mint as a garnish before serving. The soup may also be served chilled, for summer dinner parties.

Note: The spinach is included for colour, otherwise the soup would look insipidly pale. However, the quantity added can be varied. Some families may find the spinach makes the soup too bitter for their liking and that it destroys the delicate cucumber taste. Onion is also a variable ingredient.

SNACKS

Cheese Toasts

8 oz. Cheddar cheese
8 rashers lean grilled bacon
1 onion
1 teaspoon cream
1 teaspoon dry mustard
2 loaves of thickly sliced bread

Mince together the cheese, bacon and onion. Mix with the cream and mustard. Cut bread into rounds or slices without crusts. Toast on one side only. Spread mixture on other size.

Freeze in polythene bags.

To serve: Thaw at room temperature for one hour then grill under medium heat for four minutes.

Mushroom Stock for the Freezer

(This utilizes stalks, trimmings and misshapen mushrooms which cannot be sent to table as a garnish or side dish.)

Fry the mushroom pieces in butter with a squeeze of

lemon juice, salt and black pepper. Melt 2–3 oz. of butter in a pan making a roux with either flour or cornflour. Cornflour always gives better results in freezer cookery than flour, although it is a little more expensive. Add the sieved or liquidized mushroom bits and top up with ½ pint milk. When the mix has achieved the consistency of a thick paste, cook and freeze.

This mix can be diluted to form the basis of an excellent mushroom soup, used as a sauce over fish, meat or omelettes or added to stews.

Home Made Tomato Juice

Select ripe, firm tomatoes. Core and cut into sections. Place in covered pan. Crush lightly to draw out enough juice to cover the bottom of pan. Heat quickly to simmering temperature. Rub through seive. Cool quickly by placing the pan in ice water. Season with ½ teaspoon of salt to every pint of juice. Package in rigid cardboard or plastic containers, leaving head-room for expansion.

FISH

Kedgeree

1 lb. cooked and flaked smoked haddock
8 oz. Patna rice cooked and drained
3 oz. butter
Salt and pepper
1 tablespoon chopped parsley
2 hard boiled egg yolks

Melt the butter and combine with other ingredients. Pack in foil container, add lid. Freeze. Thaw in refrigerator for

3 hours, then reheat in double saucepan over boiling water or put in cool oven straight from freezer. Heat for 45 minutes.

Seafood Casserole

1 lb. lightly cooked and drained macaroni
2 tins of lobster soup
2 six oz. cans of sliced mushrooms
2 tablespoons of lemon juice
2 tablespoons of Soy sauce
2 teaspoons of celery salt
3 cups of shrimps
2 lb. of crab meat

Place the macaroni in the base of a foil dish. Heat the rest of the ingredients and pour them over the macaroni. Cool at speed. Freeze.
To serve: Defrost overnight in a refrigerator then bake for 30 minutes.

Potted Crab

2 large crabs
1 dessertspoon anchovy essence
Salt
Cayenne pepper
4 oz. butter
3 tablespoons melted butter
Pinch of mace

Take out all the crab meat, both hard and soft. Pound finely with the essence, butter and seasoning. Put in a pot and cover with melted butter. Cover with foil lid and freeze.

Fish Cakes for the Freezer

½ lb. mashed potato
½ lb. cooked white fish
2 teaspoons chopped parsley
1 oz. butter
Seasoning
Frying fat or oil
Egg
Breadcrumbs

Mix the potatoes, flaked fish and parsley together. Melt the butter in a saucepan, add the fish and potatoes, seasoning and enough egg to bind. Divide the mixture into eight portions, coat with egg and crumbs. Fry in the fat. Cool quickly, pack and freeze.
To serve at table: Thaw and reheat in one operation either in the oven or in a frying pan.

Fish cakes made to this recipe can also be stored in their raw state and fried for the first time when required for the table.

MEAT DISHES

Liver Pâté

8 oz. liver
1 small onion chopped
½ oz. fat
2 oz. bacon
4 oz. pork sausage meat
1 beaten egg
2 oz. breadcrumbs
1 teaspoonful Worcester sauce
1 tablespoon of lemon juice
1 teaspoonful celery salt
2 tablespoons red wine

Brown the liver and onion in the fat. Mince together with the bacon. Mix with the rest of the ingredients and sufficient water to create a whole of smooth consistency. Bake uncovered in loaf tin lined with foil.

Stand the tin in ice water to cool it as quickly as possible. Remove the parcel from the tin and wrap in double foil. Freeze.

To serve: Thaw in refrigerator overnight. Slice and serve with salad or as a sandwich filling.

Savoury Veal Loaf

1½ lb. minced veal
¼ lb. minced lean pork
¼ teaspoon garlic salt
6 tablespoons finely chopped onion
2 cups soft bread crumbs (8 oz.)
4 tablespoons grated cheese
2½ teaspoons salt
½ teaspoon pepper
2 beaten eggs
1 cup milk (8 fluid oz.)

Combine ingredients and shape into oblong loaf. Wrap in freezer paper and seal.

To serve: Thaw overnight in a refrigerator or place in the oven 6 to 8 hours before the oven is turned on. Bake at 350°F for 1½ hours. Makes six to eight portions.

Jugged Hare

The hare is a large animal and the flavour of jugged hare is very rich. It is unlikely that a small family will consume a whole hare in one meal or want to eat it on consecutive days. You should, therefore, prepare this dish in individual portions for immediate and future use.

The following classic recipe is ideal for freezer packs too.

1 Hare
2 oz. butter
Salt and pepper to taste
1 onion
4 cloves
1 tablespoon lemon juice
12 peppercorns
Bouquet garni
1½ pint stock

Joint the hare – and, if required for freezer storage, remove meat from bones. Fry the sections in 2 oz. of butter until brown. Place in a casserole (if required for freezer storage the casserole should be lined with foil as shown on page 76). Add the onion stuck with cloves, lemon juice, bouquet garni, stock and peppercorns. Place a tight lid on the casserole and simmer for around 2–2½ hours (hare for immediate use simmers approximately 3 hours). Freeze in containers and store.

On reheating, add one glass of red wine and some redcurrant jelly to the mix.

Cream of Rabbit

½ lb. raw rabbit
1 small egg
¼ pint white sauce
Salt and pepper to taste

Chop and mince the rabbit finely and pound until smooth. Work in the egg, white sauce and season. Press lightly into foil containers and freeze.
To serve: Remove from freezer and steam gently for 30 to

40 minutes. Serve with a hot brown sauce. Sufficient for six people.

Lamb Curry

¾ cup sliced peeled onions (5 oz.)
1 cup diced celery (8 oz.)
Clove of garlic
2 tablespoons butter or margarine
1½ cups cooked lamb cut in cubes (12 oz.)
1½ teaspoons curry powder
2 cups lamb gravy (1 pint)
1½ teaspoons salt
2 tablespoons flour
¼ cup cold water (2 fluid oz.)

Sauté the onions, celery and garlic in butter until lightly brown. Mix in the flour, lamb, curry powder and seasoning. Then add water. Cover and simmer for 30 minutes. Cool quickly by floating the pan in iced water. Pack in freezer containers. This recipe makes one quart of curry.
To serve: Heat curry in saucepan over a low heat for 45 minutes. Gives four servings.

Quick Chilli Con Carne

This recipe can be made quickly as some of the ingredients are canned. It is a dish which most men love and which is quick and easy enough for them to prepare themselves. It is also ideal for bedsitter dwellers using a mini freezer to expand their menu range.

1–2 lb. minced beef
Pepper and salt to taste
Several large onions
1 spoonful of brown sugar

1 oz. tomato purée
1 can of tomatoes
2 cans baked beans
2 level teaspoonfuls of chilli powder.

Fry the onions gently until brown. Add the sugar, seasoning and tomato purée together with the meat. Continue frying until this, too, turns brown. Turn into a saucepan, add the beans, chilli powder and canned tomatoes. Cook with the lid firmly closed for 30 to 40 minutes on a low light. Turn into individual containers or a foil basin for storage.

To serve: Put into a casserole and reheat in the oven until piping hot.

Seasoning quantities will depend on individual taste. If the mixture is preferred super-hot in the genuine South American style, top up on the chilli during the reheating stage.

Beef Stroganoff
This is another useful recipe for chain cookery. The basic buy of beef steak can be broken down into various packs of dishes, all prepared simultaneously. The time to chain cook a particular food is when it is at its cheapest or its best quality. Only then will it be a bargain buy in the bulk purchasing quantities which chain cooking demands.

2 lb. beef steak
Flour
Seasoning to taste
1 large onion
¼ lb. mushrooms
2 oz. butter
½ pint sour cream
A little brown stock

Cut the beef into small pieces. Shake in a paper bag with the pepper, salt and flour. Fry the onion in the butter until golden brown. Add the meat to the onion, together with the mushrooms. Fry until the meat is browned. Add the stock to the mixture and cook for a further fifteen minutes over a low heat. Pack into containers and freeze. The addition of sour cream gives the Strogonoff its exotic Russian flavour and this should go in at the reheat stage. When the meat is hot all through and the gravy is bubbling, but not boiling, add the sour cream. Continue reheating for a few minutes longer – again be careful not to boil the mixture.

Braised Sweetbreads

2 oz. lambs sweetbreads
½ oz. butter
½ oz. flour
¼ pint stock
Salt and pepper to taste
Bouquet garni
1 bay leaf

Wash the sweetbreads and place in a pan of water. Bring to the boil and simmer for 10 to 15 minutes. Drain off the liquid, skin and chop the sweetbreads. Make a roux of the butter and flour, add the stock and bring to the boil. Add the sweetbreads, seasoning and herbs. Simmer for a further 5 to 10 minutes. Pack into containers and freeze.

When reheating this dish, a dash of sherry or madeira will improve the flavour.

Stuffed Cabbage

1 lb. minced beef
1 oz. butter
1 teaspoon chopped parsley
$\frac{1}{2}$ teaspoonful chopped mint
2 tablespoons cooked rice
1 tablespoon sieved boiled onion
Stock from the freezer
12 medium sized cabbage leaves (looks very dramatic with
  red cabbage)
Seasoning to taste

Fry the minced meat slowly until all its liquor has been
absorbed, then add the butter and brown. Combine with
the herbs, onion, seasoning, stock and rice before cooking
for a further 5 minutes. While the mixture is cooking,
blanch the cabbage leaves. Place a spoonful of filling in the
centre of each leaf, and roll into a neat parcel. Place the
filled leaves in a casserole. Add sufficient stock to cover and
cook for around $\frac{3}{4}$ hour at 350°F (gas regulo 4). Cool and
freeze in individual portions.
To serve: Reheat in double boiler.

10

Recipes Using Commercial Ingredients as Base

The freezer owner can build up a library of recipes which utilize commercially frozen lines in interesting ways.

Here again are a few ideas which are personal favourites of mine.

SAVOURY DISHES

Beefburgers Western Style

8 oz. packet beefburgers (frozen)
16 oz. can baked beans
1 large onion sliced into rings
4 oz. mushrooms
1 oz. butter

Place the beans in an ovenproof dish. Melt the butter adding the meat, onion and mushroom, frying gently for 5 minutes. Remove the burgers and place on the beans. Cook the vegetables for a further 5 minutes then drain. Add to the burger dish and cook at 375°F (regulo 5) for 15 minutes.

Steaklet Risotto

2 oz. butter
1 large packet of frozen steaklets
1 onion finely chopped
1 clove of crushed garlic
4 oz. chopped mushrooms
2 chopped rashers of bacon
6 oz. patna rice
1 beef stock cube
1 tablespoon sherry
1 small packet frozen peas
3 skinned and chopped tomatoes
Salt and pepper to taste

Melt the butter and fry the meat. Drain and keep hot. Fry the onion and garlic until soft, add the mushrooms and the bacon and fry for 3 minutes. Add the rice and the stock cube, dissolved in 1 pint of water. Add the peas and simmer for approximately 15 minutes. Then add the tomatoes and the meat cut into quarters. Season to taste, add sherry and continue cooking until all the stock has been absorbed by the mixture.

Eggs on Cocotte

1 small packet frozen spinach
Juice of half a lemon
Pinch black pepper
6 eggs
4 oz. ham
6 tablespoons cream
Pinch of dry mustard

This dish is cooked and served in small individual portion dishes. Divide the spinach among the dishes sprinkling with lemon juice and pepper. Break an egg into each dish. Chop

the ham finely, sprinkle over the eggs and add the cream.
Sprinkle with mustard. Bake for 25 minutes in an oven at
350°F (regulo 4).

Ham and Bean Salad

2 ounces cooked and diced ham
1 small packet frozen broad beans
French dressing
Shredded lettuce

Cook packet of frozen broad beans as directed and allow
to cool. Mix all the ingredients together and garnish with
chopped parsley to serve.

VEGETABLES

Peas with Zest of Orange

2 packets of frozen peas cooked as directions on pack
Grated rind of 2 oranges

Add orange peel to the peas just before serving.

Sweet Corn and Ham Stuffing (for chicken)

1 small size packet of frozen sweet corn
1 oz. margarine
1 medium onion finely chopped
2 oz. bread crumbs
2 oz. ham (chopped)
Chopped chicken liver
1 tablespoon chopped parsley
Salt and pepper

Cook sweet corn according to the directions. Melt margarine and lightly fry onion until soft. Mix in the remaining
ingredients.

Sweet Corn Fritters

1 packet sweet corn
1 egg
2 oz. self raising flour
Pinch of salt
4 tablespoons milk
Butter or oil for frying

Cook the corn according to the packet directions. Drain
and cool. Make a batter with the egg, flour, salt and milk.
Stir in the cooled corn. Fry tablespoonfuls of the mixture
in the hot fat until golden brown and crisp on both sides.

These fritters are an essential accompaniment to southern
style fried chicken or chicken Maryland. Chicken drum-
sticks, used for both these dishes, can be purchased quick
frozen or stored after disjointing a whole fresh chicken for
chain cookery.

Sweet corn fritters are also good as a breakfast side dish
with bacon and/or sausages. They can also be eaten, like
Yorkshire Pudding, as a sweet with lemon juice and sugar
or with jam.

Fritters can be made with fresh sweet corn. Freezer store
interleaved with paper as suggested for pancakes. Reheat
also as for pancakes.

Spinach with Sour Cream

Cook frozen leaf spinach as directed on pack. Before
serving add Eden Vale sour cream (about 2 dessertspoons
to a small packet of spinach).

Leaf spinach can also be improved in flavour by the
addition of a sprinkle of nutmeg or a dash of lemon juice
before serving.

Vegetable Custard

1 large packet mixed frozen vegetables
½ pint milk
2 eggs
Salt and pepper to taste
1 tablespoon chopped parsley
3 oz. grated cheese

Cook the vegetables as directed. Heat the milk and pour over the beaten eggs adding the seasoning, parsley and grated cheese. Pour this mixture over the vegetables and bake at 350°F (or regulo 3) for 40 minutes.

This dish can be turned into a fish and vegetable casserole with the addition of cooked flaked white fish, or frozen shrimps direct from the freezer, spiced with a pinch of curry powder. Or give a meaty quality by adding any kind of diced cold meat. Ham or cooked bacon cubed is particularly successful.

Pickled Green Beans

¾ cup white vinegar
¼ cup sugar
¾ teaspoon salt
¾ teaspoon mustard seed
Large packet frozen green beans
Cup chopped onion
¼ chopped pimento

Bring the vinegar, sugar, salt and mustard seed to the boil. Add the beans and cook for 3 minutes. Add the onion and other ingredients.

Will keep for several weeks.

Beans Provençal

1 packet of sliced green beans
½ oz. butter
Salt and pepper to taste
¼ pint tomato sauce or purée (this can be from freezer
　standby supplies)
1 green pepper sliced

Cook the beans as directed and drain. Fry and toss the green pepper gently in the butter. Add the beans and tomato purée. Garnish if desired with chopped parsley, or chives and season before serving.

FISH

Crispy Cod Fries Eastern Style

1 dessertspoonful Demerara sugar
1 oz. butter
1 finely chopped onion
2 red peppers sliced
1 small can mushrooms strained from their liquor
1 small can bean sprouts strained
2 tablespoons white vinegar
1 dessertspoon mango chutney
1 packet Bird's Eye Crispy Cod Fries
1 Large Bird's Eye sliced green beans

Melt the butter and fry the onion until soft but without browning. Add the other vegetables with the exception of the beans, and add the vinegar, sugar and chutney. Heat through. Cook the beans and fish according to the packet directions. Add the beans to the mixture and place the fish on top before serving.

Mock Lobster Cocktail

1 small packet Cod Steaks
A pinch of salt, pepper, dry mustard and castor sugar
  (1 pinch of each)
2 tablespoons of cooking oil
2 oz. of frozen, peeled prawns
1 tablespoon of wine vinegar
1 dessertspoon of chopped parsley
A few lettuce leaves

Poach the steaks in a little water for about 15 minutes. Drain and cool. Mix the seasonings in a basin, add the oil and gradually add the vinegar. Flake the fish and mix with the prawns and parsley in the dressing. Serve in individual glasses with the lettuce as garnish.

Fillets of Plaice Caprice

1 small packet frozen plaice fillets
Lemon juice
2 oz. butter
Flour
Seasoning to taste
Half a banana for each fillet of plaice
Chopped parsley

Mix the flour and seasoning. Coat the fish fillets and fry in the melted butter until golden brown. Drain and keep hot. Sprinkle the banana halves with lemon juice to retain the colour and sauté. Place half a banana on each fillet and sprinkle with the parsley as a garnish.

DESSERTS

Apple Sauce Cake (can be made with home produced apple purée from the freezer)

1 thawed container (8 oz.) of sweetened apple purée
8 oz. butter
8 oz. sugar
12 oz. self raising flour
2 eggs
½ lb. mixed dried fruit
Grated lemon peel
Mixed spice

Cream the butter and sugar. Beat in 1 egg, add 4 oz. of flour, introduce the second egg and a further 4 oz. of flour. To the remaining 4 oz. of flour add the dried fruit, flavourings and last of all, the apple purée. Amalgamate all the ingredients. Bake in a well buttered shallow tin at 400°F (regulo 6) until the mix begins to rise. Then drop the temperature to 350°F (regulo 4). Bake for approximately 30 minutes.

Baked Alaska

Only freezer owners can have this dramatic sweet always on hand.

1 eight-inch sponge cake – either home made or purchased
1 tablespoonful raspberry purée made from freezer stored
    raspberries
1 ice cream brick
2 egg whites
3 oz. castor sugar

Spread the flat side of the cake with the fruit purée and put the slightly-softened ice-cream on top. Place in a flat foil container and freeze. When firm, wrap in foil.

To serve: The Alaska base, still frozen, is topped with a soft meringue, made from the stiffly-beaten egg whites and castor sugar. When placing the meringue on top of the ice-cream and cake, take care that the ice-cream layer is fully covered by the meringue. Cook in a very hot oven for 3 to 5 minutes or until the surface is a light brown. Serve immediately.

Florida Orange Jelly

1 6¼ oz. can of frozen orange juice (undiluted)
2 cans of water
½ oz. gelatine
1 peeled and cored apple (diced)

Pour the orange juice into a bowl and add the water. Dissolve the gelatine in another half can of water in a basin over hot water. Allow to cool. When cooled, add gelatine to the orange juice with the diced apple. Pour into a mould and leave until set.

11

Freezer Standbys

The clever cook will quickly learn to use her deep-freezer as a labour-saving device.

Menus can be made more nutritious and interesting if you keep a selection of 'standby' sauces and trimmings among the freezer stock. Use these to brighten other freezer-stored items or fresh bought foodstuffs. Such standbys can earn the freezer owner the reputation of being a 'perfect' host or hostess since complex, *cordon bleu* style meals can be prepared in minutes from the freezer contents. Each section of each course can be prefabricated on a day when there is time to spare for a chain cooking session.

It is a fair comment that, to the true gourmet, a freezer-stored sauce will not compare favourably with a fresh sauce made with 100 per cent fresh ingredients. However, to all but the professionally trained and expert palate the difference will be barely noticeable. Only those who refuse to have any frozen commodity in the larder or on the menu need refrain from freezing sauces on this account.

It is also said that much of the pleasure in eating comes initially from the eye appeal of the food. In this way, sauces which give colour contrast to a meal will increase the food excitement. The presentation of even as humble a food as a fish cake can be enhanced by the addition of a green-tinged parsley sauce or a pink-tinted shrimp one.

Ice-cream with melba topping will be transformed from something normally sucked in the streets to a 'Ritzy' sweet course.

Standbys will be of most value if you take advantage of your freezer to plan menus in advance. Every meal will then have its own selection of accompaniments ready to hand. Prepare standbys according to your basic menu planning. This will avoid the danger of odd pots of this and that sauce hanging around after they should have been used and having to be discarded. Often, however, the situation works entirely in reverse and the housewife can adjust her shopping to agree with her freezer standbys.

Sauce and accompaniments are best stored in individual portion containers. This allows the occasional faddy guest to have his choice, perhaps of a fancy steak butter. Such small portion pots also pack handily in odd corners of the freezer.

The recipes for standby dishes listed here represent only a small section of the total possibilities. The freezer owner will soon learn by experiment the wide scope open to her.

Freezer dishes with their own accompaniments could be the next step in commercial freezer dish presentation. The success of sauced vegetables, such as onions in white sauce, has led the big food concerns to think in terms of sauces to accompany fish and meat dishes.

Another possible development is 'reformed' food. This is already familiar in the form of fish fingers and fish sticks and burgers of all types. The method is for bulky, awkwardly shaped and bone filled food to be minced or pulped down and then reshaped into blocks, circles or fancy shapes. These will be simple to pack and space saving to store. Costs will be reduced because freezer space will be taken up by food which is 100 per cent edible. Protein-rich foods, which might otherwise be set aside as 'difficult to eat', will also become acceptable to children, invalids, the elderly and anyone wanting to have a quick snack.

'Reformed' food can also be made by the housewife who prepares freezer meals from fresh ingredients. It is especially useful for the country dweller who has large quantities of garden produce at glut seasons of the year.

Hard Sauce

This is useful to serve with steamed puddings – or in its liqueur-laden version with Christmas pudding and as the centrepiece of baked apples.

3 oz. of unsalted butter
3 oz. of brown sugar

Cream the fat and sugar until light, fluffy and coffee coloured. To this basic mix can be added a variety of flavourings:

Vanilla essence
The grated rind and 1 tablespoonful of lemon or orange
   juice
Strong black coffee
2 to 3 tablespoonfuls of fruit purée (apricot, strawberry and
   raspberry purées are especially effective for this sauce)
3–4 tablespoonfuls rum plus a teaspoon of lemon juice
   (this makes the sauce into a good equivalent of the famous
   Cumberland rum butters)
2–3 tablespoonfuls brandy

Note: Spirits can lose potency during the freezing process so more should be added to restore flavour and alcoholic content as the mixture thaws out.

Mixed Fruit Stuffing

This is a useful preparation to keep on hand for use with fatty poultry such as goose. It can also be used to give chicken an oriental slant or as a vegetarian stuffing for tomatoes.

½ cup of cooking apples, chopped
½ cup chopped, soaked prunes
¼ cup seeded raisins
¼ cup chopped oranges
Salt, brown sugar and nutmeg to taste
2 cups toasted breadcrumbs
2 tablespoonfuls melted butter
½ cup grape juice or apple juice

Mix all well together and leave for an hour before packing into freezer containers for future use.

Butterscotch Sauce

This makes a useful topping for ice cream or steamed pudding and is a favourite with children.

6 tablespoonfuls golden syrup
2 oz. brown sugar
1 tablespoonful custard powder
Juice of 1½ lemons
1 oz. butter
¾ pint water

Melt the butter, sugar and syrup in a saucepan. Mix the custard powder with the lemon juice. Remove the heated ingredients from the stove and add the water. Pour this into the custard liquor. Return this to the heat and cook gently until thickened.

This is best stored in small containers of individual

portion size. Quarter-pint wax cartons or small plastic containers are ideal.

Cream Chantilly

Fresh cream does not freeze well but when, as in this recipe, it is mixed with other ingredients it will freeze satisfactorily. To save time, use 'pre-fabricated' decorations for the tops of jellies, trifles, etc.

1 oz. icing sugar
½ pint cream
Vanilla essence

Whip the cream until thick but not clotted. Then fold in the sugar. Add the flavouring (which can be varied to taste although the classic Chantilly always utilizes vanilla). Freeze.

Apple Purée

This is a must for orchard owners who have a quantity of apples which are unsuitable for normal rack storage, but would be useful at other times of the year. The following recipe could be doubled or extended to any quantity required.

The basic mix can be used as an apple sauce for pork dishes, blended with cream to make an apple food dessert (add a touch of calvados before serving to give this an 'adult' appeal), used for apple sauce cake (see recipe on page 110), fluffed with meringue mix and baked as a soufflé (especially delicious when spiked with cinnamon). It can be used to fill an apple strudel or mixed with cut peel, spices and mixed dried fruits to fill pre-cooked pastry shells as 'English Apple Boats'.

In short, this apple purée can be the beginning of a chain cookery cycle.

To make the purée:

4 lb. cooking apples
¼ pint water
1 oz. butter
Juice and grated rind of 1 lemon
4 oz. sugar

Peel and quarter the apples. Place in a large saucepan. Add the water, butter and lemon. Cover and cook gently, agitating the pan from time to time to ensure that the mixture does not burn to the bottom. When the mixture has reached a pulpy consistency, add the sugar. Whisk until this has been absorbed and a smooth consistency is achieved.

Bread Sauce

1½ pints milk
1 small onion finely chopped
A little mace
Salt and pepper to taste
6 oz. breadcrumbs
2 oz. butter

Heat the milk in a saucepan and add the onion, mace and seasoning. Strain and then add the breadcrumbs and butter. Simmer until the mix is thick and creamy.

Pack into containers and freeze, leaving head space at the top of each carton or plastic box.

Maître d'hotel Butter

¼ lb. butter
1 tablespoonful finely chopped parsley
1 tablespoonful lemon juice
Salt and pepper to taste

Cream the butter. Add the chopped parsley and lemon juice. Season.
Serve on steaks and grills.

Garlic Butter

¼ lb. butter
1–2 cloves of garlic
1 tablespoonful finely chopped parsley
Salt and pepper to taste

Soften the butter. Add the chopped garlic cloves and pound together. Season.
Serve with grills and fish.

Basic Chicken Stock

1 chicken (can be cheap boiling fowl)
1 quart water
Bouquet garni
Salt and pepper to taste
1 stick celery
1 carrot
Some bacon rinds

Put the chicken and vegetables in a pan with the rinds and seasoning. Add the water and cover tightly. Simmer for 2 hours. Strain and cool. Skim fat off the surface. Pack in cartons and freeze.

Potted Hare

A traditional spread for toast or biscuits can be prepared from left overs of hare.

No exact quantities are given since this is an ideal recipe for using up oddments, or for chain cooking. Personal tastes will determine the precise quantity of flavourings required for the amount of meat available.

Ingredients:
Hare
Bacon
Bouquet Garni
Cloves
Pepper corns
Mace
Bay leaves
Salt and pepper to taste
Stock, made from either the hare bones or from
    chicken bones

Take some small neat, boned pieces of raw hare. Pack closely in a saucepan lined with bacon. On the top place a bouquet garni, some cloves, pepper corns, mace and bay leaves. Season with salt and pepper. Just cover with stock. Cook slowly for approximately 3 hours adding more stock as and when needed. Remove from heat and mince the hare meat and bacon. Moisten with a little more stock and season well. Place in foil containers and top each with a little clarified butter and freeze.

This recipe will provide a mix which will keep well under normal larder conditions but a freezer greatly extends the period for safe storage.

GLOSSARY OF TERMS USED IN FREEZING

Anti-Oxidant Chemical agent such as ascorbic acid (Vitamin C) which is added to sugar or syrup and controls discoloration of fruits.

Ascorbic Acid Vitamin C available from prescription pharmacists. Sometimes combined with a sugar extender and sold under a trade name.

Blanch The heating of vegetables in boiling water or steam to slow or stop the action of enzymes.

Butcher Wrap Place food near corner of wrapping paper. Fold corner over food and then fold two sides across the top. Roll package over to complete the wrap. Fasten with freezer tape.

Direct Fill Container A rigid type of container which does not need to be assembled before filling.

Dry Pack To package fruit without adding liquid.

Enzymes Chemical agents which bring about the ripening of vegetables. Their action must be stopped before freezing to prevent loss of quality, flavour and colour. (See blanch.)

Freezer Burn   Dehydration of improperly wrapped food resulting in loss of colour, flavour and texture.

Glaze   To dip frozen food quickly into iced water just above freezing temperature, so that a thin coat of ice forms over the food.

Head Space   Space left at the top of a container to provide room for expansion.

Heat Seal   To seal close with pressure from a sealing iron.

Index

In the same series

EASYMADE WINE

AND

COUNTRY DRINKS

by Mrs. Gennery Taylor

Make your own wines for less than 5p a bottle

Carrot Wine – Beetroot Wine – Dandelion Wine – Hip Wine – Lemon Wine – Marigold Wine – Prune Wine – Orange Wine – Tea Wine – These are just a few. Drinks for Children. Teas and Coffee are also here.

All so simple and economical to make. These are country recipes which have stood the test of years, and the ingredients can be easily found in fields and hedgerow.

No expensive equipment is required. All you need is a large saucepan or preserving pan, an earthenware or china bowl, and some bottles with corks.

Provide your family with inexpensive, but often potent, alcoholic beverages.

Times Literary Supplement: '. . . simpler than most, is informally written, calls for very little special apparatus . . .'

Uniform with this book